Coasts in Crisis

# Coasts in Crisis

*A Global Challenge*

Gary Griggs

UNIVERSITY OF CALIFORNIA PRESS

University of California Press, one of the most
distinguished university presses in the United States,
enriches lives around the world by advancing scholarship
in the humanities, social sciences, and natural sciences. Its
activities are supported by the UC Press Foundation and
by philanthropic contributions from individuals and
institutions. For more information, visit www.ucpress.edu.

University of California Press
Oakland, California

Library of Congress Cataloging-in-Publication Data

Names: Griggs, Gary B., author.
Title: Coasts in crisis : a global challenge / Gary Griggs.
Description: Oakland, California : University of
    California Press, [2017] | Includes index.
Identifiers: LCCN 2016051203 (print) | LCCN 2016051496
    (ebook) | ISBN 9780520293618 (cloth : alk. paper) |
    ISBN 9780520293625 (pbk. : alk. paper) | ISBN
    9780520966857 (ebook)
Subjects: LCSH: Coasts. | Coastal zone management. |
    Natural disasters. | Climatic changes. | Hazardous
    geographic environments.
Classification: LCC GB451.2 .G75 2017 (print) | LCC
    GB451.2 (ebook) | DDC 551.45/7—dc23
LC record available at https://lccn.loc.gov/2016051203

Manufactured in Hong Kong

26  25  24  23  22  21  20  19  18  17

10  9  8  7  6  5  4  3  2  1

*For my loving wife and partner, Deepika, for making it all possible*

*For my children and grandchildren, with a promise that I will do everything I can to make sure that you have healthy coasts and oceans to enjoy*

*If you are not part of the solution, you are part of the problem.*

—Eldridge Cleaver

*Courage is what it takes to stand up and speak; courage is also what it takes to sit down and listen.*

—Winston Churchill

# Contents

# Preface

I have always lived within a few minutes, or at most an hour, from a shoreline. The coast of California has been home for most of my life, but sabbatical research combined with travel, time as a Fulbright scholar, and teaching on several voyages of the Semester-at-Sea program have given me opportunities to study or observe portions of the coasts of forty-six nations. These adventures and experiences have shaped my view of the world.

My own ancestors migrated to the West Coast from New England and the Midwest, arriving on the shoreline of the Pacific Ocean a century ago. For many, the ocean has always been an edge, a destination, and a new frontier. For me personally, there has always been something very comforting and reassuring about living on the coast, with an ocean a short distance away that stretches toward the horizon and thousands of miles beyond.

From Scotland to South Africa, Portugal to Peru, Alaska to Australia, and numerous places in between, the coasts I have studied and seen share some common traits but also some substantial differences. Whether differentiated by geology, landforms, climate, beaches, population density, or intensity of development, shorelines around the world are all bathed by seawater and affected by storms, waves, and tides; they are also experiencing a rising sea while continuing to attract more people.

Coastal regions across the Earth have drawn human beings since the earliest civilizations emerged over eight thousand years ago. Despite the hazards that regularly have an impact on both large shoreline cities and small communities, migration to coastal regions steadily continues. Almost half of the planet's population now lives in what is broadly defined as the coastal zone. The impacts of these people—three billion and rising—are increasingly affecting this often-fragile meeting place of land, sea, and air, one of the most dynamic and constantly changing environments on Earth and often an area of high biological productivity and rich biodiversity.

As populations have grown, natural processes or hazards have brought coastal regions greater yearly losses and damage. Superstorm Sandy (2012), the Tohuku, Japan, earthquake and tsunami (2011), Hurricane Katrina (2005), and the Indian Ocean earthquake and tsunami (2004) are some relatively recent examples. All shorelines are also experiencing a rising sea level, which is causing coastal erosion and flooding, and perhaps a more severe future storm and wave climate. With about 150 million people around the world living within three feet of high tide and hundreds of millions more within a few more feet, future sea-level rise may be the greatest challenge human civilization has ever faced.

These dense populations, particularly the megacities, or those with over ten million people, are taking a toll on the coastal zone, affecting not only the shoreline itself but also the nearshore waters. The myriad effects include industrial, agricultural, and domestic runoff and discharge; the disposal and accumulation of plastic and other marine debris; extraction of groundwater and petroleum, leading to seawater intrusion and ground subsidence; the impacts of large port developments with their thousands of ships; overfishing and loss of habitats; and the impacts of dams, sand mining, and coastal engineering structures on sandy beaches.

The trends have been clear for some time, and the signs are all pointing in the wrong direction: more people with increased vulnerability to natural hazards and greater environmental impacts from the footprints of these increasing populations. Each of these individual hazards, risks, or issues has been studied and written about individually, but there is no source that treats the entire coastal zone as a region under threat. While this book includes more examples from California and the United States, I have attempted to cover issues and environments from a global perspective so that this book will be useful and informative for readers anywhere on the planet.

In researching and writing this book, I came to the conclusion that many of our coasts are in crisis. The oceans and their coastal zones are in the throes of a fever that is rising to increasingly dangerous levels. I chose to end each chapter with a section titled "Where Do We Go from Here?" I hope this will encourage everyone who reads this book to think about the challenges that we face and what we individually or collectively can do if we are to restore, preserve, and protect the unique and vital coastal environments.

The problems we face today as a civilization are not going to go away on their own. The future choices we make about our coasts and oceans are not just technical decisions about peripheral matters. They are decisions about who we are, what we value, what kind of world we want to live in—the world our children and grandchildren will inherit. The stakes are very high.

# Acknowledgments

I am indebted to all of the many photographers who were generous with their work and allowed me to make the topics covered in this book come to life. Some are personal friends, others are former students, but many we contacted after having seen their work shared through Creative Commons on the Web. These photographs have greatly enriched every chapter and subject. In particular, I want to acknowledge my wife and partner, Deepika Shrestha Ross, not only for her wonderful photographs but also for her encouragement throughout this project and for tracking down every image that I believed we had to include and obtaining permissions and acknowledging the photographers.

No writer is an author without a publisher, and I am indebted to the staff at the University of California Press, including Maeve Cornell-Taylor, Kate Marshall, Merrik Bush-Pirkle, Francisco Reinking, and Sheila Berg, who saw the value of this book and brought it to fruition.

# PART ONE

# Introduction to Humans and Coasts

# CHAPTER I

# Human Settlement of the Coastal Zone

Relatively early in human history it became clear that coastal regions were attractive areas for settlement. The deltas and alluvial plains adjacent to coastlines provided flat, fertile land and water that made agricultural production possible, the mild climate made life easier and more comfortable, and the coastal waters provided access to the sea. Over time, trade and commerce would develop. There were some long steps for early humans, however, from the grasslands of Africa to the Nile Delta or the fertile crescent of the Middle East, but over thousands of years these areas were gradually settled. The earliest civilizations were preceded by the domestication of plants and animals and the development of agriculture, which date to 10,000 years ago plus or minus a few thousand years. The beginnings of agriculture took place over several thousand years at nearly the same time in several different areas, including Egypt, the Fertile Crescent of the Middle East, India, China, Middle America, and parts of Europe.

## THE ROLE OF CLIMATE CHANGE AND SEA-LEVEL RISE IN COASTAL SETTLEMENT

There are many ideas and hypotheses as to what triggered or led to the transition from our hunter-gatherer ancestors to farmers, among them the climate changes that took place when the last ice age ended and the modern Holocene epoch began (usually dated at about 11,700 years

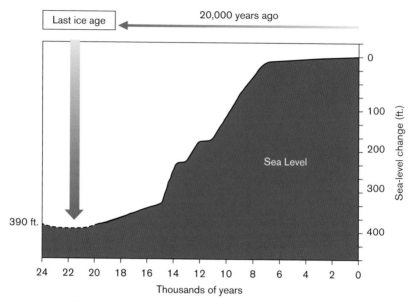

FIGURE I.I. Global sea level has been rising for the past 20,000 years, since the last ice age ended.

ago). Much of the Earth became warmer and drier, which favored annual plants that died back but produced seeds (grains) or tubers that could be cultivated, harvested, and stored for later consumption. This was a huge step forward for humanity.

Another argument has been proposed for a connection between the early development of agriculture and the stabilization of sea-level rise. As the last ice age came to a close about 18,000 years ago, glaciers and ice sheets gradually began to melt and seawater warmed, increasing the volume of the ocean and thereby raising sea level. There was a period of fairly rapid warming and associated sea-level rise between about 18,000 and 7,000 years ago (figure 1.1). During this period, sea level rose on average nearly half an inch per year or about 45 inches per century. Within this approximately 11,000-year period of warming, there were also what are believed to have been meltwater pulses when glaciers retreated very rapidly, causing sea level to rise even faster. During these intervals, the oceans were rising at nearly an inch a year or over six feet per century. In low-relief and low-lying deltas or coastal plains, a few feet of sea-level rise could move the shoreline landward thousands of feet or more, so these areas would not have supported permanent agriculture or settlements.

About 7,000 years ago, however, climate change slowed and the rate of sea-level rise declined dramatically. Until about a century ago global sea levels were fairly stable, rising only about 0.04 inch per year, equivalent to about 4 inches per century. This created some stability and the opportunity for early humans to begin to occupy and settle the fertile coastal environments and, along with the warmer climate, begin to cultivate crops. The earliest evidence of civilization appeared within about a thousand years of the cessation of sea-level rise, although there is still considerable debate over the most important contributing factors.

During the Neolithic (~12,000 to 7,000 years ago), a number of population groups began to abandon hunting and gathering and for the first time started to settle in villages, domesticating animals and tending crops. Settlements in that era tended to be in inland areas, in the valleys, foothills, and mountains, and burials at that time usually lacked indications of social classes or distinctions. Within about a thousand years of sea-level stabilization, however, a transition seems to have taken place, with people beginning to migrate to coastal areas, where communities began to develop with significant increases in population, as well as burials that show the existence of social classes. These communities also began to construct monumental architecture, indicative of societies with large labor forces (think about the Great Pyramids of Egypt). These concentrations of people and labor forces required a large and dependable food supply, and the stabilization of sea level allowed that in several ways. Rivers delivered soil, nutrients, and organic matter to the more stable coastal plains and deltas, allowing for essentially continuous agricultural production. In addition, the nearshore marine environment stabilized, with highly productive wetlands, estuaries, intertidal zones, and reefs, which soon were discovered to be important year-round food sources to complement what was grown on the adjacent land. Archaeological records from sites around the world from this time period show that fish and shellfish, as well as marine mammals, were part of the emerging food supplies made possible by this stable and productive coastal environment. The development of irrigation canals and the construction of fishponds further enhanced this new coastal margin productivity. Early civilizations gradually developed and expanded, although it would be a few thousand years more before the world's coastal regions began to be recognized as sites of considerable value to human settlement and cities began to develop and expand. In addition to important settlement sites, coasts or shorelines likely provided routes for migration to new areas.

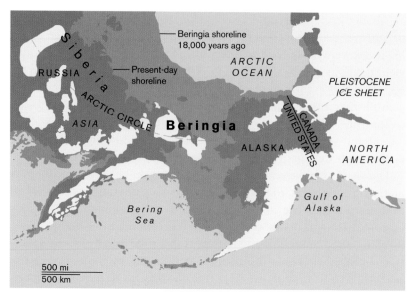

FIGURE 1.2. Route of early human migration to the Americas from Asia across the Bering Land Bridge.

Recent discoveries at the Monte Verde archaeological site in southern Chile has confirmed this as the oldest known human settlement in the Americas discovered to date and provides additional evidence in support of the theory that one early migration route followed the Pacific coast of the Americas more than 15,000 years ago, from one end of the Americas (the Bering Strait) to the other (Patagonia). In fact, to the surprise and initial disbelief of many anthropologists and archaeologists, this migration appears to have taken place remarkably quickly. The Monte Verde site shows the existence of a group of people living along the beaches and banks of sand and gravel of a small stream about 14,800 years ago.

The dominant theory since the 1930s suggests that the ancestors of Native Americans crossed the Bering Strait about 15,000 years ago, when sea level was about 350 feet lower than today (figure 1.2). They were likely following herds of large game, and these early visitors then quickly spread out through North and then South America. The recent analysis of genetic data, however, based on the rate at which mutations occur, indicates that the first Native Americans were isolated from their Asian ancestors for thousands of years before dispersing through the Americas. They likely departed from their Siberian relatives about 25,000 years ago and entered Beringia, a former landmass that now encompasses the far eastern area of Russia, the shallow parts of the Bering Sea, and western Alaska. Evidence also indicates that this area was likely drier grassland where large mammals like wooly mammoths, giant ground sloths, steppe bison, musk ox, and caribou grazed. Steller sea lions were also present along the shoreline during this time. The presence of an abundant food supply from the ocean and the land, as well as woody plants that could be used for fires and shelters, may have provided a reasonable place to settle down for a while, perhaps until ice melt opened up migration paths into North America.

Based on the number of sites in the Americas where evidence of human habitation has now been uncovered, stretching from Alaska to Chile, and despite some archaeological arm wrestling over what counts for concrete evidence of human presence and what does not, it seems that humans arrived in the Americas at least 15,500 or 16,000 years ago. Pushing the arrival date back introduced another problem to be solved, however. Based on sea levels and ice coverage 15,000 years ago, there does not appear to have been an ice-free highway for these early visitors to follow into North America.

The lack of a convenient or even passable route on dry, ice-free land from Beringia introduced the idea that perhaps the first Americans did not walk here at all but came in small boats and followed the coastline south, perhaps sustaining themselves with the abundant marine life of the nearly continuous kelp highway that borders North America's west coast. This idea was first proposed in the late 1950s after very old human bones were discovered on Santa Rosa Island, off the coast of Santa Barbara, California. The bones were later determined to be male, named the Arlington Springs Man, after the discovery site. Significantly, the bones were later dated at 13,000 years, making these the oldest human remains found anywhere in the Americas at the time. Although at that time sea level was about 225 feet lower than today, Santa Rosa Island was still separated from the rest of California by about 5 miles of ocean. Unless this early man was an Olympic swimmer, he and his friends and family must have crossed the deep Santa Barbara channel by boat. Pygmy mammoth bones of the same approximate age were also unearthed on Santa Rosa Island, indicating coexistence with this early human and representing a long swim for a small pachyderm. While there seems to be no question that the earliest Americans dined on marine animals, it is still not clear if this nearshore kelp highway was a route used by the earliest Americans in some sort of primitive boat.

## COASTAL SETTLEMENTS AND CITIES

Between the Arlington Springs Man and the Monte Verde site in Chile, it appears that at least some of the earliest human inhabitants of the Americas found the coastal areas to their liking. About eight thousand years later, the early Mediterranean civilizations, including the Egyptians and then later the Phoenicians, Greeks, and Romans, were some of the first peoples to realize the advantages of coastal areas. The development of small coastal vessels and then larger ships allowed these groups to benefit from easy access to the sea, and most of the major cities of those civilizations were seaports or had ocean access. Some eastern Mediterranean cities had natural harbors, which facilitated marine trade and commerce and the offshore expansion of fishing, which diversified diets.

A thousand or so years later, the Vikings and their predecessors adapted successfully to very different coastal environments and mastered the craft of boat building and seafaring as they colonized the Baltic and North Seas. They also reached America and established colonies five hundred years before Columbus arrived.

Each of these early coastal civilizations inhabited distinct regions or environments, from the low-relief and constantly shifting but very fertile Nile Delta to the natural or constructed harbors of the eastern Mediterranean to the deep fjords of Scandinavia, but each group benefited from the presence of the ocean. The climate of these coastal regions was nearly always more moderate than inland areas, which experienced greater extremes in temperature. In most cases, the adjacent ocean also provided a supply of protein for the growing populations. Harbors, whether natural or manmade, became new centers of trade and commerce, simply because ships allowed for the first large-scale transport of goods to other areas. They also became important for defense and military activities, and ships allowed for the development of the first navies and more effective ways of transporting soldiers and implementing invasions or battles.

As civilization advanced and populations grew, coastal regions became progressively more important, and many of Europe's large cities developed along or near coastlines as ports and centers of commerce. Athens, Venice, Rome, Lisbon, London, Amsterdam, Copenhagen, and Stockholm come to mind. As Europeans settled the Americas in the seventeenth and eighteenth centuries, many major cities were founded on the coast, around the Great Lakes, or along navigable rivers, among them New York, Boston, Chicago, Toronto, Washington, DC, Detroit,

Cleveland, and Montreal. As the west coast of the United States and Canada was settled, San Diego, Los Angeles, San Francisco, Portland, Seattle, and Vancouver emerged as coastal cities and ports.

But whether in Europe, the Americas, or Asia, the earliest settlements and the development of communities and then cities on the coast were in many cases related initially to the ability to develop agriculture on floodplains and deltas, harvest fish and shellfish from the adjacent ocean, or become centers for maritime trade. Over time, the added advantages of access to the sea for commerce and defense became equally or more important.

It was not until the late 1800s and early 1900s, however, when railroads and steamships made travel easier and faster, shorter workdays allowed for leisure time, and an urban middle class with disposable income emerged, that there developed oceanside accommodations and resorts for leisure pursuits. This began an entirely new era, with increasing numbers of people heading to the coast for recreation and relaxation, whether the coast of the Mediterranean, Brighton or Seabright in England, the Greek Isles, Australia's Gold Coast, San Diego in Southern California, or the Hawaiian Islands. People came to the coast initially for short stays and then more recently for retirement, with progressively greater impacts.

## HUMAN IMPACTS ON COASTAL ENVIRONMENTS

Each of the chapters that follow focuses on a specific issue relative to greater human occupancy and use of coastal regions. The chapters in part 1 describe the natural processes and hazards that affect coastal regions around the world and the people who live and work there. Those in part 2 discuss the major impacts that human activities are increasingly having on coasts globally.

The footprints of the early humans who occupied and settled the coast were quite modest in comparison to today's impacts. Dune vegetation was sometimes grazed by domesticated animals, which resulted in destabilization, forests were cut down, and sediment loads to the shoreline were increased, leading to coastal accretion or outbuilding and damage to coral reefs in tropical latitudes. Waste discharge was minimal, however, and fishing pressures were initially low. These impacts were all recoverable because they were generally on a small and local scale relative to what is possible with the population numbers, machines, and technology that developed over the past 150 years or so.

The coastline began to change, however, in response to natural processes as well as human activities. The ancient Greeks and Romans were very capable engineers and built ports and harbors, along with their monumental architecture. Today many of these early ports are filled with sediment and are several miles inland from the present shoreline as a result of a thousand or more years of sedimentation.

There was also the progressive awareness that living at the edge of the ocean presented significant hazards. Tsunamis have taken large death tolls historically, in Japan, in the Indonesia archipelago, and even on the Mediterranean coastlines. Cyclones, hurricanes, and typhoons have also taken their toll over the years throughout South Asia. Despite these risks, people have continued to be drawn to coastal areas. Cities have grown along with exposure to natural hazards and the impacts of the expanding populations on coastal environments and natural systems.

## THE COAST AS A DESTINATION

The attraction of the coast as a vacation or holiday destination exploded after World War II for a number of reasons but built on what had begun fifty or more years earlier. Widespread automobile ownership brought access to the coastline within the reach of most people, regardless of income. Campgrounds and caravan parks replaced farmland and grazing land. Vacation resorts multiplied, hotels expanded, and new attractions, such as marinas and golf courses for those who could afford them, were added to draw even more people. Many former sleepy fishing villages along the Mediterranean coast of Spain, France, and Italy, if they had beaches, became summer resorts for the sun-craving people of northern Europe. High-rise condominiums and apartments were constructed by the thousands to accommodate these seasonal visitors, which took their toll on the social and cultural fabric of these former towns but also provided new types of employment and increased economic activity. Oceanfront promenades and boardwalks often replaced hauled-out fishing boats and drying nets.

All of this changed the character of coastal towns, although the sun and warm ocean waters continued to draw people: the Costa Brava, Costa Blanca, Costa del Sol, Costa Verde, Costa de la Luz, to name a few in Spain (figure 1.3). Florida is a lot like Spain's Mediterranean coast, drawing people from New York and New Jersey first to vacation and then, often, to relocate permanently. The beaches in Florida have

FIGURE 1.3. Intensive shoreline development at Peniscola along the Mediterranean coast of Spain. (Photo: D. Shrestha Ross © 2015)

always been magnets, but they are getting narrower in many areas due to the effects of jetty construction at inlets. Miami is also building higher and higher, as if trying to outpace the increasing rate of sea-level rise.

## COASTAL POPULATION EXPANSION

Regardless of how we define "the coast" or "the coastal zone," it has become substantially more crowded globally than inland regions in recent decades, and all indications are that this trend will continue. The number of people around the world occupying coastal regions is difficult to determine, and the figures vary according to different sources and criteria: approximately 3 billion people living within 125 miles, 3 billion people living within 100 miles, or 3 billion people living within 60 miles, you can take your pick. Give or take a few hundred million, this is a big number, and it's growing, producing an expanding crisis at the coast.

In the United States in 2010, of the three million total square miles of land area (excluding Alaska), less than 20 percent (512,971 sq. mi.) is in *coastal watershed counties* (counties containing watersheds that drain to the coast). But 52 percent, or 163.8 million of the U.S. population of

313 million, live in these counties. Less than 10 percent of the nation's land area (275,351 sq. mi.) is in *coastal shoreline counties* (those that border on a coast), but 39 percent, or 123.3 million people, live in these counties.

Not surprisingly, following a global trend, population densities in U.S. coastal regions are high and getting higher, simply because more people are crowding into the same desirable areas. In 2010 the average population density of the United States as a whole was 105 persons per square mile. For coastal watershed counties, average density was three times as high, 319 persons per square mile. For coastal shoreline counties, the density was 446 persons per square mile, or over four times as high. Nationally, population density grew to 36 persons per square mile in the forty years between 1970 and 2010. In contrast, coastal watershed counties added 99 persons per square mile, while coastal shoreline counties grew by 125 persons per square mile. Many of these new migrants to coastal counties are older or retired, with the percentage increase in people sixty-five or older being significantly higher in coastal states; South Carolina had a 443 percent increase in seniors in the period 1970–2010; Hawai'i, 340 percent; Virginia, 248 percent; and Florida, 208 percent.

To meet the housing needs of these migrants, building is continuing in coastal areas. New building permits were being issued at the rate of 1,355 per day in coastal shoreline counties across the country between 2000 and 2010.

## URBAN EXPANSION AND THE RISE OF COASTAL MEGACITIES

Major coastal cities continued to grow during the twentieth century and into the twenty-first to become some of the largest urban areas on the planet. Eight of the world's largest cities are on the coast, and their cumulative population is just over 200 million. Eighteen of the largest twenty-five cities are also coastal and combined total 357 million people, almost 5 percent of the world's total population (figure 1.4). The list of coastal megacities (usually defined as having over 10 million people) includes Tokyo-Yokohama, New York City–Newark, Shanghai, Mumbai, Karachi, Buenos Aires, Los Angeles, Rio de Janeiro, Manila, Osaka-Kobe, Lagos, Istanbul, Guangzhou, Shenzhen, and Jakarta. The number of megacities and their populations are both projected to increase substantially in the decades ahead.

New York
20m
24m (+20%)

Karachi
14m
20m (+43%)

Beijing
16m
23m (+44%)

Tokyo
37m
39m (+5%)

Rio de Janeiro
12m
14m (+17%)

Los Angeles
13m
16m (+23%)

Shanghai
20m
28m (+40%)

Calcutta
14m
19m (+36%)

Manila
12m
16m (+33%)

São Paulo
20m
23m (+15%)

2011
2025

Buenos Aires
14m
16m (+14%)

Mumbai
20m
27m (+35%)

Dhaka
15m
23m (+53%)

FIGURE 1.4. Populations of global coastal megacities in millions of people in 2011 and projections for 2025 (with % increase).

These coastal megacities clearly continue to offer what are perceived to be better economic and social opportunities than whatever rural areas where many people formerly lived, but they are not paradise by any stretch. These very high concentrations of people require large land areas for housing, as well as food, water, and energy to sustain themselves, which all come with costs or impacts. And the expectations of a better life and improved living conditions are typically not met due to high poverty and unemployment rates as well as social rejection in many of these areas. The added problems of a lack of sanitation and waste management infrastructure and inadequate housing conditions have also led to serious problems of human and environmental health.

In many large coastal cities where available land or topography permits limited growth, the shoreline has often been extended seaward with fill or reclamation of one sort or another to provide more buildable land area. Singapore is one of the best examples of this; its land area has been increased an astonishing 22 percent by importing rock, earth, and sand, and more expansion is under way. Hong Kong, which was historically landlocked, severely limiting growth or expansion, is another example. Land reclamation in Hong Kong actually began in the late 1860s. Hong Kong International Airport and Hong Kong Disneyland were both built on land reclaimed from the sea. Considerable fill has taken place in prime locations on both sides of Victoria Harbor, and a

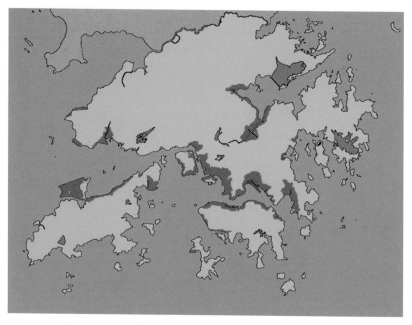

FIGURE 1.5. The shoreline of Hong Kong has been extended seaward over 150 years of land reclamation (areas in gray have been added; those in red are planned).

number of entirely new towns have been built in recent decades on reclaimed land, among them Tuen Mun, Tai Po, West Kowloon, Kwun Tong, and Tseung Kwan O (figure 1.5).

Mumbai, India, was originally an archipelago of seven separate islands that have all been joined by reclamation over several hundred years. Macau has increased its land area 170 percent, or 6.6 square miles. Ninety-six square miles of Tokyo Bay, Japan, have been reclaimed, including the construction of Odaiba Island. Kobe, Japan, has added 8.9 square miles, creating several intensively developed "port islands." The Netherlands may be the best-known example of land acquisition, with about one-sixth of the nation (2,700 sq. mi.) reclaimed from marshes, swamps, lakes, and the sea. As a result of sea-level rise, the Dutch are starting to give some land back.

The wetlands and marshes around the margins of San Francisco Bay began to be filled over 150 years ago, at the time of the California gold rush. The filling continued with little control or oversight until 1970, when the Bay Conservation Development Commission (BCDC) was established to bring the haphazard and unplanned filling under control.

FIGURE 1.6. Approximately one-third of the original extent of San Francisco Bay has been filled (areas shaded in brown) for development. (Image: Exobyte via Wikimedia, in the public domain)

In the meantime, however, about one-third of the entire bay had been filled to make space for housing developments, commercial and industrial buildings, as well as two international airports and baseball parks (figure 1.6). In the case of San Francisco Bay, which is underlain by many feet of soft and compressible mud, subsidence and differential settlement and cracking of buildings and streets created major problems in areas of artificial fill. Shoreline land reclaimed around the world has the potential to subside or settle depending on what the underlying material consists of, the nature of the fill, and the engineering and method of emplacement.

Around San Francisco Bay, the fill and the underlying sediments also are highly susceptible to liquefaction during strong seismic shaking if not engineered carefully, and the Bay Area is surrounded by active faults. Damage during the great 1906 San Francisco earthquake and again in the 1989 Loma Prieta earthquake, which was centered 75 miles from downtown San Francisco in the Santa Cruz Mountains, was greatest in areas underlain by fill. The largest number of fatalities in the 1989 earthquake (forty-two of sixty-three) occurred during the collapse of a section of freeway in the Oakland waterfront area that had been built on deep water-saturated sediment and fill.

Careful engineering, including replacing the poorest soils or materials with engineered fill, can reduce the hazards of settlement, subsidence, and liquefaction during seismic shaking. Surcharge or excess loading for extended periods prior to construction can accelerate compaction and settlement and reduce subsequent subsidence. Supporting structures on deep piles or caissons that extend to firm material or bedrock is another common approach for large buildings. There will always be pressure to expand the boundaries of megacities, and shoreline land reclamation has been the most common approach historically. There are many areas in tropical waters where important habitats (coral reefs in the case of Singapore and Hong Kong, for example) have been covered over with fill or destroyed in the past. These losses are not recoverable; once they are lost, they are gone. Coastal planning and controls on future reclamation need to be an essential part of any government's future policies and coastal protection measures. The San Francisco BCDC provides an excellent model for what can be accomplished with political will and governmental interagency cooperation.

## VULNERABILITY OF LARGE CITIES TO COASTAL HAZARDS

Global sea-level rise and short-term elevated coastal ocean levels from hurricanes, typhoons, cyclones, El Niño events, and tsunamis are discussed in chapters 2, 3, 4, and 5. Hurricane Katrina, the 2011 Tohoku, Japan, earthquake and tsunami, Superstorm Sandy (2012), and Typhoon Haiyan (2013) in the Philippines are all tragic examples of short-term events, which are usually followed by government aid for reconstruc-

FIGURE 1.7. Intensive high-rise development along the shoreline of Miami Beach, Florida. (Photo: D. Shrestha Ross © 2015)

tion in the same low-lying hazardous locations. These short-term events will continue to present the greatest risks to coastal cities around the world over at least the next several decades. However, over the longer term, sea-level rise combined with future storms or extreme events may well present the greatest challenge that human civilization has ever faced. More people are moving to and living in coastal areas, and many of the planet's biggest cities were built along shorelines very close to sea level, originally at a time when sea-level rise was not a concern. And there are those today who still do not believe that sea level is rising or that it is of concern. We don't get to vote on sea-level rise. It's happening; the rate is increasing, and 95 percent of climate scientists believe that human activity is the greatest driver.

By 2050, according to the United Nations, more than 6 billion people are expected to live in cities. More than half of the world's largest cities are ports, and they will face unique challenges due to their low-lying elevations and increasing populations. Over 130 port cities around the planet are at increasing risk from rising sea level, combined with severe storm-surge flooding, damage from high storm winds, and, in some places, local land subsidence. It is safe to say that none of these cities were planned with sea-level rise in mind. In many cases, construction is proceeding as if sea level were standing still (figure 1.7). Miami,

Florida, is one of the best examples. In 2016 from Miami to West Palm Beach 417 condominium towers (with over 50,000 individual units) were being built, and not one of them was taking sea-level rise into account. Poorly planned development often puts more people in vulnerable areas, increasing future risk.

The Organization of Economic Cooperation and Development (OECD) has determined that in 2015 the twenty global port cities with the highest vulnerability had 25.2 million people and $2.2 trillion in assets exposed to a 100-year storm. By 2070, however, with additional sea-level rise and projected population growth, these same twenty port cities will have approximately 88 million people at risk, along with a projected $27 trillion in assets. And this is just twenty of 130 global port cities. The number one city on the list in terms of exposed assets is Miami, with $416 billion at risk in 2015, which is projected to rise to $3.51 trillion by 2070.

## HUMAN IMPACTS ON THE COASTAL OCEAN

Not only are natural hazards taking their toll on the large coastal concentrations of people, whether in megacities or in smaller cities and towns, but the three billion people living in coastal areas are having their effects as well. Chapters 6 through 17 discuss the wide range of human impacts on the coastal ocean. The by-products of civilization are diverse, widespread, and complex. Waste discharge, whether domestic, industrial, or agricultural, is damaging to the coastal ocean and its life in multiple ways, and the more people, the greater the volumes of discharge. Marine life and the habitats they occupy, from temperate to tropical latitudes, have all been affected. While renewable energy from the ocean is an admirable and necessary long-term goal, we have made only modest progress. We are still heavily dependent globally on fossil fuels, and their extraction, transport, combustion, and by-products are taking their toll, from oil spills to greenhouse gas emission, with their effects on global climate and ocean acidification. Dams in our coastal watersheds and sand mining along the shoreline have damaged our beaches, which are critical to tourism and our own recreation.

As was stated in a popular comic strip decades ago, "We have met the enemy and he is us." The Earth cannot support 7.5 billion people in the lifestyle the wealthier nations have grown accustomed to. The coastal zones of the world, more than any other environment or geography, are where the impacts of civilization have been concentrated and are becoming most obvious. The chapters that follow offer an explanation and a status report on these multiple stressors and their effects. In each case and for every issue, the chapters conclude with thoughts, perspectives, and recommendations for the future—how we can begin to reduce our impacts on the coastal zone and improve the health of the coastal environment for all of us to enjoy.

# PART TWO

# Natural Processes and Hazards Affecting Coastal Regions

# CHAPTER 2

# Coastal Tectonics and Hazards

## SUBDUCTION ZONE EARTHQUAKES

While more and more people are migrating to coastal regions around the world, these geographies present many different types of geologic hazards. Most coasts around the Pacific Rim are areas of tectonic collisions. It is the colliding of massive lithospheric plates that gives rise to very large subduction zone earthquakes and their associated tsunamis, as well as volcanic activity, although the latter can also occur in mid-ocean regions such as Hawai'i, far from plate boundaries.

The 1960s and 1970s witnessed a revolution in scientific thinking about the history and evolution of the Earth. Much of the initial evidence for the developing theories and concepts came from the exploration of the ocean basins, which began in earnest in the 1950s with the availability of surplus navy vessels for oceanographic research following World War II. The discovery of a world-encircling, volcanically active, undersea ocean ridge system and a nearly continuous circle of deep ocean trenches surrounding the Pacific Ocean with associated chains of active volcanoes (the "Ring of Fire") led to the concept of seafloor spreading in 1963. The subsequent recognition that a relatively narrow band of earthquakes followed the ridge and trench system (figure 2.1) led to the development of the theory of plate tectonics in 1968. New, exciting, and somewhat controversial nearly fifty years ago, this theory now forms part of the basic geologic history of the Earth that many students learn about in grade school today.

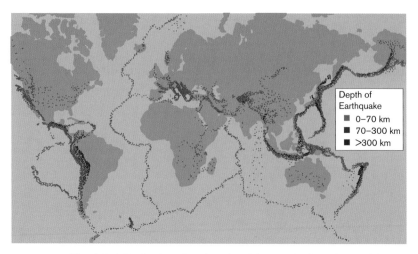

FIGURE 2.1. The global distribution of earthquakes delineates the boundaries of the Earth's tectonic plates.

Plate tectonics provides a comprehensive framework and explanation for the origin of the large-scale features of the Earth—the mountain ranges, volcanoes, trenches, earthquakes, and faults—and how they fit together. Coastlines around the Pacific Rim and the northern Mediterranean are like IMAX theaters of geologic change. If you can't see some interesting geology outside your window, you can usually drive a few miles in any direction and you will. These coasts are for the most part geologically active and prone to movement at a moment's notice, up, down, or sideways.

The coasts on either side of the Atlantic are quite different, however. Many of them are much older geologically, more subdued in their topography, and relatively stable tectonically. They do not shift around a lot. When was the last time you heard about a big earthquake in England, France, or Brazil? Residents of these countries certainly have hazards to contend with, but those tend to be things coming from the ocean, atmospheric phenomena like hurricanes or severe storms. Much of the geologic history along the edges of the Atlantic Ocean took place millions of years ago, and things are generally quite peaceful today.

The interior of the Earth is hot, partially molten in places and therefore weak and somewhat mobile. This hot fluid material slowly rises, owing to its lower density, much like a pot of water heated on the stove. Where the fluid material reaches the surface, much of this along the ocean floor, the seafloor cracks open and new oceanic crust is created. This process has created a globe-encircling volcanic mountain range that passes through all of the world's ocean basins: the interconnected Mid-Atlantic Ridge, the Mid-Indian Ridge, and the East Pacific Rise, in outward appearance much like the seam on a baseball. The hot material beneath the seafloor spreads out in both directions (a process known as sea floor spreading), carrying along the Earth's crust, including the ocean floor as well as the continents (which combined to form the lithosphere), as it moves along like a giant conveyor belt.

We now recognize that the Earth's surface consists of seven large and eight smaller, somewhat thin (about sixty miles deep), rigid lithospheric plates that move around relative to one another, driven in part by the flow of hot fluid material in the asthenosphere beneath them, as well as by the pull of gravity where the edges of some oceanic plates descend or are subducted at deep-sea trenches. Much like large icebergs floating in the ocean, these plates are in constant, although slow, motion (an inch or two per year). At their edges these plates may diverge, collide, or grind past one another. The interaction at the edges of these huge plates leads to tectonically active regions where most of the world's earthquakes occur (see figure 2.1). In fact, it was the recognition of the concentrated global distribution of these zones of earthquakes that led seismologists to first recognize and sketch the boundaries of the major plates.

It is the collision of a thick, lower-density continental plate with a thin and denser oceanic plate that generates the planet's largest and most damaging earthquakes and that can have devastating impacts on coastal regions. When these two types of plates collide, something has to give, and it is the denser oceanic plate that is overcome and forced downward into the mantle. This usually creates a deep-sea trench where the seafloor is dragged downward (figure 2.2). Trenches encircle virtually the entire Pacific Basin, extending northward from the Peru-Chile Trench along the entire west coast of the Americas all the way to the Aleutian Trench and then southward down the opposite side of the Pacific Ocean from the Kamchatka and Kurile Trenches all the way to the Hikurangi Trench near New Zealand.

The Pacific does not have a monopoly on trenches and plate collisions, but it does contain most of them. The Indian Ocean has the Sunda Trench, which produced the huge magnitude 9.1 Andaman Islands earthquake and tsunami on December 26, 2004, creating death and destruction around the northern Indian Ocean. The Caribbean has the Puerto Rico Trench, the Atlantic has the South Sandwich Trench between the tip of South America and Antarctica, and the Mediterranean Sea has the Hellenic Trench or Trough.

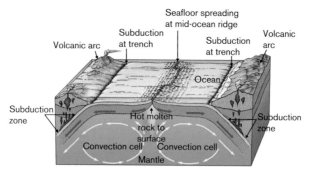

FIGURE 2.2. A cross section of the Earth's interior illustrating convection cells within the mantle that drive the motion of tectonic plates.

These seafloor depressions form the surface expression of a subduction zone, created as the descending oceanic plate pulls down the edge of the overlying continental plate and seafloor. This upper surface of the oceanic plate grinds slowly against the lower surface of the continental plate, and the friction keeps these two plates locked for centuries until the stress can no longer be accommodated. The eventual rupture or slip between the two plates creates earthquakes, often very large ones; in fact the greatest earthquakes on the planet typically occur at these subduction zones or trenches. Sixteen of the seventeen largest earthquakes recorded since 1900 have all occurred at these features, twelve of these in trenches surrounding the Pacific Basin and the other four at trenches in the Indian Ocean.

Why are these earthquakes so large and devastating to coastal areas? The size, or magnitude, of an earthquake is determined primarily by the surface area of the rupture and the amount of slip or separation between the two plates. Because subduction zones extend for several hundred miles into the mantle and trenches may be hundreds of miles long, these rupture areas can be huge. The ten largest earthquakes since 1900, which ranged from 8.7 to 9.5 in magnitude, were all at subduction zones. Two of these, Japan's magnitude 9.0 on March 11, 2011 (at the Japan Trench), and Sumatra, Indonesia's magnitude 9.1 on the day after Christmas 2004 (at the Sunda Trench), are still fresh in the minds of many living in those areas.

Like many geologic hazards, floods for example, earthquakes tend to occur at frequencies inversely proportional to their magnitudes. In other words, the really large events do not occur nearly as often as smaller events. Globally since 1900, on average, there has been only a single earthquake of magnitude 8 or larger each year, although there have been about fifteen events of magnitude 7.0 to 7.9. Based on observations since 1990, there were about 134 earthquakes between 6.0 and 6.9 magnitude annually and about ten times as many shocks of magnitude 5.0 to 5.9 (~1,319 events).

There are some important relationships to be kept in mind between earthquakes of different magnitudes. Each magnitude change (e.g., from 5.0 to 6.0) equates to a tenfold increase in ground motion and about thirty-two times more energy release. So a magnitude 7.0 shock is not just a little bit bigger than a magnitude 6.0; it is a whole lot bigger and has the potential to produce a whole lot more damage.

Because of the subduction zones and associated volcanoes that surround the Pacific Ocean, virtually every coastal area around this so-called Ring of Fire is subject to very large earthquakes, not every year, but often enough to be of concern, and impossible to predict. Any earthquake above magnitude 6.0 or so is potentially devastating, and the global death toll from these large earthquakes averages about 20,000 annually. While any earthquake of magnitude 6.0 or larger can be damaging, it is those that occur in regions with high population densities, weak foundation conditions, and poor construction practices or materials that have produced the greatest number of fatalities.

A striking example of the importance of these site-specific factors is the contrast between fatalities from two nearly identical magnitude events, the 1989 6.9 Loma Prieta earthquake in Central California and the 2010 7.0 Haiti earthquake. Total fatalities in California were 63, while the "official" death toll in Haiti was 316,000, although there is uncertainty about this number, and it may have been considerably lower. The major difference between these two events was the type of construction. Most of the deaths in Haiti were a result of the collapse of unreinforced masonry buildings. No one died in California in 1989 in a single-family dwelling, yet the earthquake was nearly the same magnitude.

Since 1900 there have been 125 individual earthquakes around the world that have led to at least 1,000 deaths, and about one-third of these have been at subduction zones or have affected coastal areas, including the two very large twenty-first-century events: Sumatra in 2004 (~228,000 fatalities) and Japan in 2011 (~21,000 fatalities).

In addition to the ground shaking that accompanies any earthquake, which is the process that affects the largest area and often produces the greatest amount of damage and loss of life, fault rupture, slope failure or landslides, ground settlement, and liquefaction usually add to the

FIGURE 2.3. The library at the university in Tangshan, China, was extensively damaged by a major earthquake in 1976 but was left as a reminder of the event, which may have killed as many as 250,000 people, making it the deadliest earthquake of the twentieth century. (Photo: Gary Griggs © 2003)

destruction. Distance to the earthquake epicenter, the size of the earthquake, the nature of the foundation materials, and the type and quality of construction have repeatedly been shown to be the key factors in determining damage, injuries, and fatalities. Earthquakes rarely kill people, but collapsing buildings frequently do (figure 2.3). In the January 2010 Haiti earthquake, over 97,000 houses were destroyed. In the May 2008 Sichuan, China, earthquake (magnitude 7.9), an estimated 5.36 million buildings collapsed and over 87,000 people died.

In general, when passing from dense crystalline rock (e.g., granite), through less dense sedimentary rock (sandstone or shale) into unconsolidated and finally water-saturated alluvial materials, seismic waves tend to become amplified, and shaking is both more severe and of longer duration. Cities located on river floodplains or alluvial river valleys have repeatedly suffered far greater earthquake damage than those built on firm bedrock. In virtually any coastal region around the world, these differences in underlying geologic materials can be identified and delineated so that decisions on future building or infrastructure and engineering criteria do not have to be made naively or in a vacuum. We know what we can expect in the future.

It has become painfully evident that type of construction has an extremely important influence on how well structures perform during strong seismic shaking. Many of the most destructive earthquakes in terms of loss of life during the past century took place in regions where the dominant construction materials were unreinforced masonry, concrete, brick, or adobe. These occurred in countries like China, Pakistan, Iran, Turkey, Japan, Italy, Peru, and Guatemala, and in Nepal in 2015, where thousands of buildings collapsed and over 8,500 people died.

Unfortunately, lots of buildings in coastal areas around the world were constructed without an adequate understanding of seismic forces and used substandard construction methods. Materials (unreinforced masonry, adobe, brick) that have repeatedly performed very poorly during large earthquakes are still being used because wood or steel is unavailable or unaffordable. New construction in areas of recent historic earthquakes needs to be carefully designed, engineered, and inspected to reduce future fatalities. We need to learn lessons from past disasters, and implementing modern building codes and construction practices has to become a priority. While this may seem logical and obvious, there are many reasons that this is not routinely done: cost savings, short disaster memories, corruption in the construction and inspection process, or low priority, such that earthquake deaths in many areas continue to be very high.

## EARTHQUAKES AFFECTING COASTAL ZONES ALONG TRANSFORM PLATE BOUNDARIES

Subduction zones are not the only type of plate boundaries that threaten coastal cities. The San Andreas Fault in California, the Alpine Fault in New Zealand, and the North Anatolian Fault in Turkey are examples of areas where plates are sliding past one another and where moderate to large earthquakes occur somewhat regularly. All these faults are quite long, and all pass under populated areas where the potential for damage, injury, and loss of life is high.

These three faults are not simple confined breaks that you can walk over in a few steps but, due to the huge size of the plates that are grinding beside one another and the complexity of the geology, can be many miles wide. As it passes through the greater San Francisco Bay Area, the broad boundary between the North American and Pacific Plates stretches from the shoreline across the San Francisco Peninsula and well into the East Bay, where branches of the San Andreas Fault system underlie the cities of Oakland and Berkeley (figure 2.4). The great 1906 San Francisco earthquake, a magnitude 7.8 event, which leveled and burned about 90 percent of the city and apparently killed at least 3,400 people, took place along this plate boundary. San Francisco is surrounded by water on three sides and is very prone to earthquakes. It has been labeled by some who do not live there as "the city that waits to die." It is not the only coastal city facing this dilemma, however.

In Southern California, while the actual San Andreas Fault passes east of the greater Los Angeles area and its 19 million residents, the plate boundary is a zone nearly two hundred miles wide and stretches from as far as one hundred miles offshore to the arid desert, with a number of active branches of this fault system passing under the city. There has been a long history of large and damaging earthquakes in the region—Santa Barbara, 6.8 magnitude (1925); Long Beach, 6.3 magnitude (1933); San Fernando, 6.4 magnitude (1971); and Northridge, 6.7 magnitude (1994)— all with significant loss of life and major damage. Because the actual rupture area of these faults is so much less than those of large subduction zone earthquakes, however, the earthquakes are somewhat smaller; but because of population densities, damage and death tolls can still be significant.

The Alpine Fault extends through the North Island of New Zealand from Cook Strait on the south to the Bay of Plenty on the north and forms major topographic features, including the Wellington harbor, which is a down-dropped block along the fault. The Wellington section of the fault presents a major hazard as it passes beneath the capital city and is crossed by a number of bridges, roads, and pipelines. Over 75 percent of Wellington's 200,000 people live within 6 miles of the fault. Field research indicates that the length of time between large earthquakes along the fault has varied from less than 100 to over 285 years. The last most recent earthquake, however, occurred in about 1717, when over 200 miles of the fault ruptured. The 300 years since that event have been the longest period in recent centuries without a major

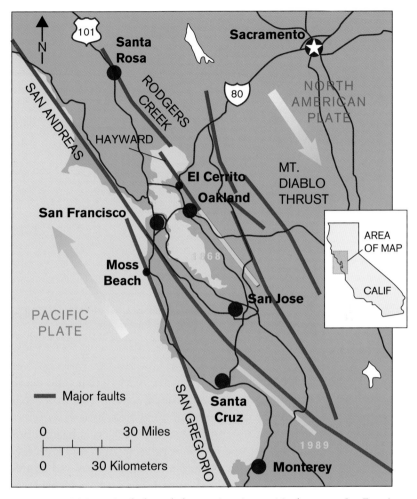

FIGURE 2.4. Major active faults and plate motions (arrows) in the greater San Francisco Bay Area. Segments of faults with recent ruptures and dates are delineated in yellow.

earthquake. This would be an appropriate time for Wellington to get prepared if it is not already. The magnitude 7.8 earthquake of November 14, 2016, centered just across Cook Strait on the northern South Island near Kaikoura, damaged Wellington extensively, and several large buildings had to be demolished as a result.

The North Anatolian Fault extends over 900 miles across northern Turkey to the Aegean Sea (similar in length to California's San Andreas Fault; figure 2.5) and lies about 12 miles south of Turkey's biggest city,

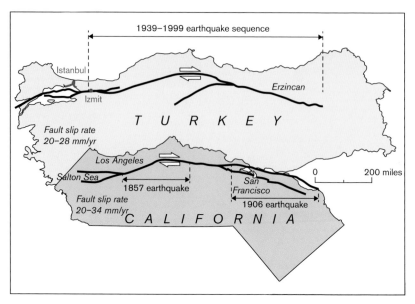

FIGURE 2.5. Comparison of the San Andreas Fault, California, and the North Anatolian Fault, Turkey, showing length, slip rate, and extent of ground rupture during earthquakes over the past 200 years. (Modified from and courtesy of Ross Stein and Serkan Bozkurt)

Istanbul. Since the magnitude 7.9 earthquake in 1939 that took nearly 33,000 lives, there have been seven subsequent earthquakes over 7.0 in magnitude. Each of these earthquakes ruptured areas progressively farther to the west, suggesting to seismologists that each event has a role in triggering the next. The last two events in 1999, in Izmit and Duzce, of magnitude 7.6 and 7.2 respectively, took over 18,000 lives and injured about 44,000 people. This earthquake migration is not believed to be complete; the next major event could likely be close to Istanbul, just 12 miles from the Anatolian Fault and with a population of 14 million.

## TSUNAMIS

Large subduction zone earthquakes generate tsunamis, or seismic sea waves. Although these events are often referred to as "tidal waves," they have absolutely nothing to do with the tides. Japan's long history with these phenomena is evidenced by the word *tsunami*. It is a combination of two Japanese characters, *tsu*, meaning "harbor," and *nami*, meaning "wave," because for centuries this is where these large waves were witnessed.

Why do tsunamis occur? When the frictional resistance between the descending oceanic plate and the overlying continental plate is finally exceeded by the cumulative motion between the two converging plates, rupture occurs. The edge of the upper plate, which has been pulled down for decades or centuries by accumulated stress, is finally released and rebounds upward, displacing huge volumes of ocean water along the way. It is this displaced water at the sea surface that generates a series of waves that move outward from the region affected by the sea-floor displacement. Most of the wave energy travels at right angles to the trend of the trench or subduction zone along which the rupture occurred, or both directly seaward and toward the shoreline.

On November 1, 1755, a very large offshore earthquake (estimated at ~8.7 magnitude), likely centered along the complex tectonic boundary between the African and Eurasian plates, killed an estimated 70,000 people in Lisbon, Portugal, making it the worst natural disaster in recorded European history. In addition to the collapse of many of the city's buildings, a tsunami came up the Tagus River within about forty minutes of the initial shock. Many Lisbon residents had fled to the river to escape collapsing and burning buildings only to be swept away by the tsunami. The waves breached 23-foot-high seawalls in Morocco and were still 10 to 15 feet high when they struck the Caribbean island of Antigua after crossing 3,300 miles of the Atlantic Ocean. This deadly event initiated the birth of the discipline of seismology and ultimately led to Lisbon being the city where some of the world's first earthquake-resistant buildings were constructed.

The March 27, 1964, magnitude 9.2 Good Friday Alaskan earthquake, the second largest ever recorded, generated a tsunami that moved directly onshore and also offshore, perpendicular to the axis of the Aleutian Trench. The waves moving onshore reached elevations of about 200 feet above sea level and, along with some local tsunamis created by submarine and subaerial landslides, were responsible for the deaths of 106 people. The tsunami moving seaward damaged the coasts of British Columbia, Washington, Oregon, and California (11 people died in Crescent City, 1,600 miles away), as well as Hawai'i and Japan.

The tsunami generated by the 2004 magnitude 9.0 Sumatra/Andaman earthquake was responsible for more fatalities than any other tsunami in history, affecting the coastlines of fourteen countries around the Indian Ocean. The sudden uplift of the seafloor 6 feet or more during the earthquake displaced massive amounts of ocean water that resulted in a tsunami that reached the adjacent Sumatra coastline within

FIGURE 2.6. The 2004 Sumatra tsunami washed as far as a mile and a half inland. (Photo: Noel Gavin licensed under CC BY 2.0 via Flickr)

fifteen to twenty minutes. Waves reached maximum elevations of up to 100 feet above sea level and washed about a mile and a half inland, leaving death and destruction in their path (figures 2.6 and 2.7). The great majority of the approximately 228,000 deaths were related to the tsunami. Indonesia was the hardest-hit nation (over 131,000 fatalities), followed by Sri Lanka (35,322), India (12,405), and Thailand (5,395). The tsunami also drowned people on the opposite side of the Indian Ocean along the shorelines of Somalia, Tanzania, Yemen, and Kenya. In South Africa, 4,800 miles across the Indian Ocean, 8 people died due to abnormally high sea levels and the tsunami waves.

Although there was a time lag between the earthquake and the tsunami, nearly all of the victims were completely surprised, in part because there was no recent history of tsunamis in the Indian Ocean and in part because there was no warning system in place at the time. Following the loss of life, injuries, and destruction from this tragic event, a tsunami warning system was established in the area and became functional in 2006.

The 2011 Tohoku, Japan, magnitude 9.0 earthquake was accompanied by 23 to 33 feet of seafloor uplift along the Japan Trench, which generated a huge tsunami that propagated both directly onshore and offshore across the Pacific Ocean. Along the eastern coast of Japan, the

FIGURE 2.7. Most buildings in Banda Aceh, Indonesia, were completely destroyed by the 2004 tsunami, although this large mosque remained mostly intact. (Photo: Guy Gelfenbaum, U.S. Geological Survey)

tsunami reached maximum elevations of 133 feet, traveled up to 6 miles inland, flooding an estimated 217 square miles, and devastated entire towns with thousands of fatalities (figures 2.8 and 2.9). Over 90 percent of the nearly 21,000 earthquake-related deaths were due directly to the tsunami.

The waves spread out across the Pacific Ocean and led to elevated water levels and damage from Alaska to Chile (figure 2.10). Between the Pacific Tsunami Warning Center in Hawai'i and the U.S. National Tsunami Warning Center, watches and warnings went out across the Pacific Basin and along the West Coast from Alaska to California. With very accurate advance warnings, evacuations were successfully carried out, although the 8-foot-high tsunami surges damaged several harbors and a number of boats in California and Oregon. Along the coasts of Peru and Chile, the surge was high enough to damage several hundred low-lying houses. Because of the warnings, however, there were only two reported fatalities outside of Japan, one in Indonesia and one in California, the latter a photographer standing on the beach to photograph the arriving waves who did not take the warning seriously.

FIGURE 2.8. The 2011 Tohoku, Japan, tsunami carried large boats inland and in this case left a boat stranded on top of a building. (Photo: Stephen Vaughan © 2011)

FIGURE 2.9. The Tohoku, Japan, tsunami devastated more than 40 miles of coastline in Iwate prefecture in northeastern Japan including Yamada Town. (Photo: Katherine Mueller, IFRC, NOAA/NGDC)

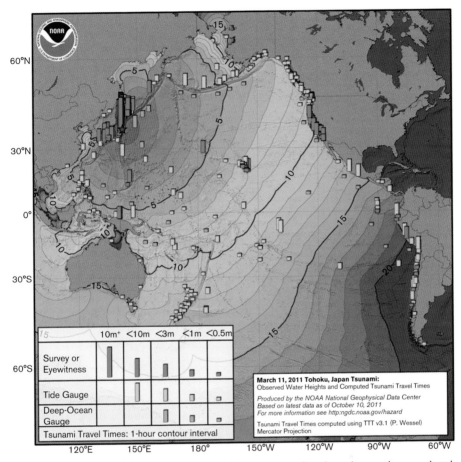

FIGURE 2.10. Heights of the 2011 Japan tsunami were greatest directly onshore and were reduced in height as the waves propagated completely across the Pacific Ocean but increased as they washed up on the shoreline of the Americas. (Map courtesy of NOAA)

Although Japan has a long history of earthquakes and tsunamis as a result of being a very old civilization and its location adjacent to a plate boundary and subduction zone (the Japan Trench), damage and casualties were high due to the unexpectedly large surge of water and high coastal population densities. A number of floodwalls and monuments marking high water from past tsunamis were overtopped.

Perhaps the most publicized and devastating example was the overtopping of a 33-foot-high seawall that was built to protect the backup power and cooling system for the Fukushima Daiichi nuclear power plant. The tsunami flooded and disabled the backup power system, which led to the loss of cooling water, explosions, and the meltdown of three of the plant's six reactors (see figure 9.9). This led to the uncontrolled release of radiation into the adjacent ocean for years. The government of Japan believes that total economic losses associated with the Tohoku earthquake may reach $300 billion, making it the costliest natural disaster in human history.

Over three hundred years ago, on the evening of January 27, 1700, a series of large waves suddenly inundated the east coast of Japan. The arrival and impacts, meticulously recorded, were a mystery at the time. While tsunamis were not new to the coast of Japan, in this case there was no nearby earthquake preceding the arrival of the tsunami, which was the expected pattern. These records were literally buried for almost three centuries until geologists doing fieldwork along the coasts of Oregon and Washington discovered evidence that the coastal area had subsided many years before, killing vegetation from submergence in salt water. Ghost forests, consisting of Sitka spruce and red cedar trees and stumps, were still preserved along the shoreline and could be dated using dendrochronology, the close examination of their annual growth rings. This analysis indicated that the trees had died between 1699 and 1700.

Careful study of sediments preserved in a number of estuaries and tidal inlets along the Pacific Northwest coast of the United States over the past several decades revealed buried vegetation covered with layers of clean beach sand, which seemed oddly out of place in these muddy, marshy environments. Dating of these deposits matched with the tree ring data and also was consistent with oral histories of native people living in the coastal area at the time, which mentioned severe shaking of the ground and flooding. The ghost forests, the buried vegetation and sediment record, the oral histories of native peoples of the Pacific Northwest, combined with the ancient tsunami in Japan, referred to now as the "Orphan Tsunami" because there was no local parent earthquake, all pointed to a very large earthquake and tsunami in January 1700 that took place on the underlying Cascadia subduction zone (figure 2.11). The plate motion and earthquake led to the subsidence of the coastline, drowning trees and marsh vegetation, which was followed by the tsunami that transported beach sand inland over a mile and a half along tidal channels, covering low-lying marsh areas.

Sediments first cored from the seafloor by geologic oceanographers at Oregon State University in the 1960s, and more recently as well, preserved the deep-sea record of these same events. Great earthquakes disturbed the accumulated sediments perched offshore on the continental

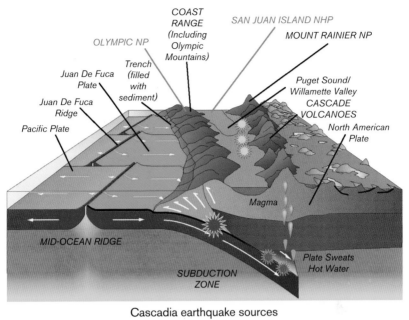

Cascadia earthquake sources

| | Source | Affected area | Max. size | Recurrence |
|---|---|---|---|---|
| ✴ | Subduction Zone | West. WA, OR, CA | M 9 | 500–600 years (1700) |
| ✴ | Deep Juan De Fuca Plate | West. WA, OR | M 7+ | 30–50 years (1949, 1965, 2001) |
| ✴ | Crustal faults | WA, OR, CA | M7+ | hundreds of years? (CE 900, 1872) |

FIGURE 2.11. The Cascadia subduction zone extends from Northern California to Vancouver Island and has produced very large earthquakes every 300 to 500 years. (Courtesy R.J. Lillie and National Park Service)

slope, which flowed downslope as turbidity currents or mudflows, leaving the sandy and muddy deposits behind on the deep-sea floor as a record of these massive events. The seafloor sediments preserved the history of at least nineteen such coupled events over the past ten thousand years, a massive (likely magnitude 9.0) earthquake followed by a large tsunami, or a catastrophic event every 525 years on average.

Today coastal communities from Southern California to Washington's Olympic Peninsula are increasingly becoming aware of their future tsunami risks through this relatively recent "paleotsunami" research, the investigation of the geologic record of ancient tsunamis. The research methods pioneered here are now being employed in other subduction zone settings around the world, which are the impact zones of historic

FIGURE 2.12. Tsunami warning signs have now been installed in low-lying coastal communities along the coasts of California, Oregon, and Washington. (Photo: Gary Griggs © 2012)

and future large earthquakes and tsunamis. If we can recognize and date prehistoric tsunami deposits and their inland extent, determine the average recurrence intervals, or return times, between these large events, as well as how much time has elapsed since the last major event, we then have the ability to more accurately assess which areas are at the highest risks and how often or soon. This is not simple, but it is doable where the sediment or vegetation records have been preserved and can be identified and carefully studied.

Along the U.S. West Coast, these paleotsunami records have led to the posting of tsunami warning signs along low-lying vulnerable sections of coastlines (figure 2.12), which warn people to evacuate

FIGURE 2.13. A mural on the wall of a commercial building in Crescent City, California, commemorates the 1964 tsunami that killed eleven people and destroyed much of the downtown area. (Photo: Gary Griggs © 2008)

or move quickly to higher ground if a large earthquake occurs or if they are otherwise notified. Crescent City, on the Northern California coast, was extensively damaged by the tsunami generated by the 1964 Alaskan earthquake, with the loss of eleven lives and the destruction of twenty-nine blocks of the business district (figure 2.13). Today the area most heavily damaged in 1964 and closest to the coastline has been preserved as open space and converted to a park to reduce future damage. Identifying and delineating those areas that tsunamis have reached in the past, their inland extent and elevation, is an important starting point in planning for potential future inundation and reducing damage. While the passage of time can begin to erase the memory of loss of life and destruction from earlier events, we need to accept the reality that plate motion, subduction zone earthquakes, and tsunamis are a fact of life in certain geographic regions. We know where these regions are and that these events will occur again. It is our collective responsibility to mark the areas of historic inundation, post warning and evacuation signs, and ensure that our warning systems are in place and working.

## VOLCANOES: ISLAND ARCS, OCEANIC RIDGES, AND HOT SPOTS

The locations of active volcanoes around the planet can be directly connected to the concept of plate tectonics and our understanding of how the interior of the Earth works. Looking around the world, there are three distinct types or areas of volcanism, all of which can potentially affect coastal regions, including ocean ridge volcanism, chains of volcanoes related to subduction zones, and hot spots.

Along the Mid-Atlantic Ridge, East Pacific Rise, and Mid-Indian Ridge spreading centers, where plates are being pulled or spread apart, basaltic magma has risen from deep within the mantle and has cooled and congealed to form an undersea volcanic mountain range about 40,000 miles in length. Larger outpourings of lava can form individual peaks, which rise in places to pierce the ocean surface in places like Iceland, the Galápagos, and the Azores, some still active and others dormant.

Iceland straddles the Mid-Atlantic Ridge and is literally being pulled apart by seafloor spreading (figure 2.14). The island consists entirely of volcanic rock, and since the time of earliest human occupation, about 900 C.E., eighteen volcanoes have erupted. The eruption of Skapter Jokul in 1783 was a national disaster for Iceland. One-fifth of the population (ten thousand people) died from the direct effects. One-half of the cattle, three-fourths of the horses, and four-fifths of the sheep population were killed as well. Six hundred miles away in Scotland, the volcanic ash destroyed crops.

In 1973 the eruption of Kirkjufell on the small island of Heimey off the southern coast of Iceland erupted and sent lava flows into the fishing port of Vestmannaeyjar, threatening to close off the harbor entrance and destroy the island's main economy. In one of the largest examples ever of trying to halt or divert lava flows, huge pumps were brought in to pump seawater onto the advancing lava front in an effort to cool and solidify it, thereby forming a basaltic dam. The efforts were ultimately successful but not before the volcano had destroyed or severely damaged over eight hundred buildings and forced 5,300 islanders to temporarily abandon their homes and the island itself.

In April 2010 the eruption of Eyjafjallajokull in Iceland spread volcanic ash over extensive areas of northern Europe, which disrupted air traffic over twenty countries and affected as many as 10 million air travelers. Volcanic ash, or tephra, is like ground glass and can damage jet engines. In this case the ash cloud rose to heights of about 30,000 feet

FIGURE 2.14. Iceland sits directly on the Mid-Atlantic Ridge and is slowly being split apart by volcanic activity. This view is directly down the rift zone where spreading is occurring. (Photo: D. Shrestha Ross © 2014)

and entered the northern hemisphere jet stream, which carried it into the paths of hundreds of commercial flights, which had to be rerouted or canceled.

Another type of volcano occurs parallel to trenches or subduction zones where the descending lithosphere is being consumed at depth as it is carried into the mantle. As the lithosphere gets deeper and hotter, superheated gases are released that partially melt the overlying mantle, and this molten material, now less dense than the surrounding fluid, starts to rise. Eventually, through a complex subsurface plumbing system, this magma works its way to the Earth's surface, above the deeper part of the subduction zone, and erupts as a volcano or, over time, a line or chain of volcanoes.

The volcanoes known as the Ring of Fire that nearly encircle the Pacific Ocean occur landward of the oceanic trenches and have been produced by magma generated during subduction. About 80 percent of the Earth's active volcanoes—those stretching from Chile up through South and Central America, the Cascades (which extend from California to British Columbia), the Aleutian chain and Kamchatka, down through Japan, and then south through the Philippines and New Zealand—are all related to the subduction of this giant Pacific Plate system. The long line of volcanoes along the west coast of Sumatra and Indonesia, including Krakatoa and Tambora, are related to the Java Trench. In the Caribbean Sea, the twelve active volcanoes of the West Indies (including Mount Pelée, Soufrière, and Soufrière Hills) are a result of subduction along the Puerto Rico Trench. The Mediterranean has its own trench and subduction zone where the African Plate is being subducted beneath the Eurasian Plate, and that has generated the eruptions of Stromboli, Vesuvius, and Etna in Italy and Thera or Santorini in Greece, in addition to many others.

While these are not all necessarily coastal hazards, depending on the topography of specific regions, the eruption of individual volcanoes has been devastating for a number of heavily populated coastal areas downslope from these active peaks. The list of the most destructive and tragic historic examples includes Mount Pelée and the city of St. Pierre in the West Indies (1902), Mount Vesuvius and the city of Pompeii in Italy (79 C.E.), Krakatoa (1883) and Tambora (1815) in Indonesia, and Thera in the eastern Mediterranean (~1613 B.C.E.).

Mount Pelée erupted violently on May 8, 1902, after weeks of warning. The volcano itself is about 400,000 years old but had lain relatively dormant, although its geologic history indicates an eruption about every 750 years for the past 5,000 years. Although today the earthquakes and release of sulfurous gas that preceded the May 1902 eruption would probably have been cause for evacuation, there was an election scheduled for May 11. The government in power, due to instability and the threat of being voted out, actually brought in soldiers to prevent people from leaving the city. In the days immediately preceding the disastrous eruption, a volcanic mudflow had inundated a sugar plantation, killing twenty-three people. The same mudflow created a tsunami when it hit the sea, which inundated the waterfront area, killing an additional sixty-eight people. At 7:50 A.M. on May 8, as people were going to church services, a black cloud raced down the side of the volcano, moving at an estimated speed of 350 miles an hour. The port city of St. Pierre, which was immediately downslope from the volcano, was engulfed in a matter of minutes by a *nuée ardente*, or glowing cloud, a rapidly moving mixture of very hot gaseous and solid particles. All the houses were unroofed or otherwise demolished (figure 2.15). The force of the blast tore apart concrete and stone walls up to 3 feet thick. The nuée ardente moved out over the ocean at the port, destroying eighteen ships and killing their crews as well. There were reportedly only two survivors from the city of about 29,000 inhabitants. The warnings were

FIGURE 2.15. View of the city of St. Pierre on the Caribbean island of Martinique after the August 30, 1902, eruption of Mount Pelée showing nearly complete destruction of buildings and scorched hillsides. (Photo: U.S. Library of Congress Prints and Photographs Division)

there, but politics intervened and the city and its people were lost. St. Pierre has subsequently been rebuilt, in the shadow of Mount Pelée, awaiting the next eruption.

The most destructive and deadliest volcanic eruption in human history occurred in August 1883, when Krakatoa, in what was then the Dutch East Indies, erupted. This volcano owes it origin to subduction of the Indo-Australian Plate beneath the Eurasian Plate. Indonesia contains over 130 active volcanoes, more than any other nation on Earth. There had been months of precursory activity at Krakatoa: earthquakes, explosions, and steam and ash eruptions. On August 27, four massive explosions took place. The largest was heard over 1,800 miles away in Perth, Western Australia. Each of these events also generated large tsunamis, perhaps reaching heights of over 100 feet along the surrounding coasts. The accompanying eruption of more volcanic ash and pyroclastic flows of lava and volcanic rock, combined with the tsunamis, took over 36,000 lives, with the tsunamis accounting for over 90 percent of the deaths. After these explosions, most of the island of Krakatoa had disappeared. As a result of the huge amounts of ash and sulfur emitted, average temperatures in the northern hemisphere in the summer were as much as 2.2°F lower than normal and weather patterns were abnormal for some years.

There is one more group of volcanoes that does not fit into either of the other two categories and that includes some of the world's largest eruptions and individual volcanoes. These can be found on continents and in oceans, in the middle of plates and at ocean ridges. The Hawaiian Islands, Yellowstone, the Azores and Canary Islands, Easter Island and Samoa, perhaps forty in all, are the result of intraplate volcanism. These are believed to be due to hot spots where molten plumes of magma from relatively stationary sources of heat within the Earth's mantle rise to the surface and erupt. Hot spots are not located only at plate boundaries, and why this is the case is not yet well understood.

In the Hawaiian Islands, the slow movement of the Pacific Plate to the northwest over a hot spot for millions of years has produced a chain of volcanoes that extends 2,200 miles from the Big Island on the southeast to beyond Midway Island on the northwest where there is a bend in the volcanic chain. The Emperor Seamount chain extends an additional 1,500 miles to the Aleutian Trench. Moving from Hawai'i, which is active and may even be erupting as you read this, up the Hawaiian chain to the northwest, each of the islands is sequentially older and no longer active. As the Pacific Plate, moving northwest about 4 inches per year, carried each island off the hot spot, the deep subsurface system connecting them to the mantle plume was shut off, and the island cooled and then slowly subsided. Moving to the northwest, Maui is 0.8 to 1.3 million years old, Moloka'i is 1.3 to 1.8 million years old, O'ahu is 2.3 to 3.3 million years old, and Kaua'i is 3.8 to 5.6 million years old (figure 2.16). Midway, now an atoll, was erupting 25 million years ago.

The Big Island of Hawai'i has formed in the past 700,000 years and erupts frequently along the East Rift zone of Kilauea (figure 2.17), which along with Moana Kea and Moana Loa have played important roles in building the island over time. Mauna Loa is the largest individual volcano on Earth, and with its depth to the deep-sea floor and elevation above sea level its total relief, at 31,000 feet, surpasses Mount Everest. This volcano has erupted thirty-three times since 1843, expanding the island in the process. A number of eruptions occurred in the twentieth century. The most recent event, in 1984, produced nearly

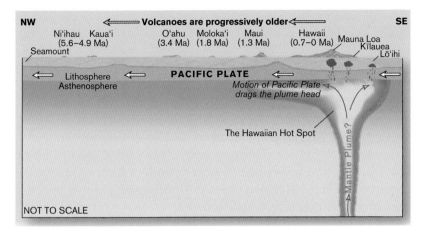

FIGURE 2.16. The Hawaiian chain of volcanic islands and seamounts was formed by the motion of the Pacific Plate over a hot spot in the mantle (ages of islands in millions of years [Ma]).

FIGURE 2.17. Kilauea on the Big Island of Hawai'i is a site of active volcanism. (Photo: Gary Griggs © 2014)

170 million cubic yards of lava and came within 4 miles of the island's biggest city, Hilo. A seamount named Lo'ihi is slowly growing on the seafloor southeast of Kilauea, and while about 3,000 feet below sea level, it may become the next island in the chain.

The Hawaiian Islands are examples of shield volcanoes, which are characterized by the eruption of very fluid magmas that can flow considerable distances and inundate large areas. Everything ever constructed on the Big Island has been built on lava of some previous eruption, whether highways, homes, or hotels. The highest risks to future inundation by lava flows occur in areas downslope from eruption centers and where lava has flowed recently. There is a long history of development and infrastructure built too close to active rift zones where destruction has repeatedly occurred and where future development needs to be avoided.

## THOUGHTS FOR THE FUTURE: WHERE DO WE GO FROM HERE?

The tectonic setting of any particular coastline is literally set in stone, and there is nothing we can do to change this. The overall approach to dealing with the large-scale geologic hazards that these areas face, whether earthquakes, tsunamis, or volcanic eruptions, includes recognizing their presence, evaluating the risks they pose, and taking every action feasible to reduce the impacts and losses from future events.

With each of these hazards, the question is not if a major event will happen again but when and how often, and herein is a huge dilemma for any community, government agency, or nation. Using the Cascadia subduction zone, which extends about 600 miles from Cape Mendocino in Northern California to Vancouver Island off British Columbia, as an example, how do you begin to respond to a very large and catastrophic event that only occurs every three hundred to five hundred years? We can make very general predictions about where we can expect these massive earthquakes and tsunamis to occur in the future (virtually anywhere around the rim of the Pacific Ocean, for example). With continued geologic and historical research we also may eventually be able to determine how often, on average, very large subduction zone earthquakes and their associated tsunamis may occur, and even perhaps when the last event took place. We still have no method, however, for predicting exactly when the next large event will happen. Volcanoes are different in this respect, in that there are usually a number of precursors to a major eruption. We can't prevent the eruption, but at least warnings and evacuations can take place, which can greatly reduce fatalities and injuries.

We now have a good network of tsunami detection, prediction, and warning systems in place, and these are very useful for coasts hundreds or thousands of miles away from the generating subduction zone or trench on the opposite side of the ocean. The system worked very well during the 2011 Tokohu, Japan, earthquake, when tsunami arrival times were predicted quite accurately along virtually the entire west coast of the Americas. Some damage still occurred, but loss of life was almost zero. The greatest risk, however, will always be in the areas directly onshore from the subduction zone, where there is precious little warning time, perhaps 15 to 30 minutes. Fatalities in Sumatra in 2004 and in Japan in 2011 made this painfully evident, and we can expect this to be the case in the future. The partial answer here lies in education and warning systems, including tsunami evacuation signs and routes, or in access to or construction of elevated structures where people can find safety.

Cannon Beach, a small, low-lying and tsunami-prone community along the Oregon coast, has plans to build what appears to be the first tsunami-resistant building, elevated 15 feet, that would provide an evacuation zone for up to 1,500 city residents and also serve as a new city hall. A number of other coastal communities in Oregon have developed tsunami warning videos that are being used in schools and community groups to prepare residents. And virtually all of the low-lying coastal communities along the U.S. West Coast now have tsunami warning signs in place.

The physical damage from these large events to towns or cities and their infrastructure, critical facilities, homes, and other structures is a much larger concern, however, as the disaster at the Fukushima Daiichi nuclear plant made clear. While a protective seawall had been constructed to protect the emergency generators in the event of a tsunami, the wave overtopped the wall and shut down the generators. It is unlikely that large cities will immediately be relocated, although with a gradually rising sea level this will happen at some future time. In the meantime, and for the foreseeable future, we need to be very conservative in both the construction of new infrastructure or buildings and how we deal with existing structures or facilities.

A road, a parking lot, and even homes present one level of risk, but a power plant, a sewage treatment plant, an airport, and similar high-use, expensive, and critical facilities are a different challenge altogether.

Most important, we should not be building any more such facilities in areas that would likely be exposed to hazards such as tsunamis during their lifetimes. The record of inundation or damage areas from past events should begin to provide the location constraints needed to make these decisions. How we deal with existing structures or facilities is clearly bound up in politics and economics, and there isn't a lot of money lying around to rebuild or replace expensive infrastructure. After a disaster, when losses are large, or before new structures are contemplated or planned is certainly one time when these decisions on long-term risk avoidance should be made.

# CHAPTER 3

# Tropical Cyclones, Hurricanes, and Typhoons

On November 8, 2013, Typhoon Haiyan, also known as Super Typhoon Yolanda, the strongest tropical cyclone ever to make landfall based on wind velocities, cut a devastating swath across the central Philippines. The storm strength was equivalent to a Category 5 hurricane (the highest level of intensity) with sustained wind speeds at landfall of 195 miles per hour (mph), the highest ever recorded, and gusts up to 235 mph. Nearly 13 million people were affected, 13 percent of the nation's entire population. There were at least 6,300 fatalities and 28,700 injuries. Because of the lightweight construction materials and the extreme wind velocities, over 281,000 houses were reported destroyed, with 1.9 million people displaced (figure 3.1).

This super typhoon had been closely tracked for six days across the western Pacific as its strength increased before making landfall. As the storm intensified, evacuation warnings were given, reaching the highest level of warning, indicating very high winds. Although wind speeds were extreme, the major cause of damage and loss of life appears to have been storm surge. Water reached 17 feet above sea level at the center of the low-lying and most damaged area, Tacloban City, where the terminal building at the airport was destroyed when the surge reached the second story. The entire first floor of the Tacloban City Convention Center, which was serving as an evacuation shelter, was submerged by storm surge. Many people in the building were surprised by the fast rising waters and subsequently drowned or were injured.

FIGURE 3.1. Damage from Typhoon Haiyan (Yolanda) in the streets of Tacloban, Leyte Island, the Philippines, November 14, 2013. (Photo: Eoghan Rice, Trócaire/Caritas licensed under CC BY 2.0 via Flickr)

The most immediate threats to survivors of this typhoon, in order of urgency, were lack of safe drinking water, no shelter, untreated injuries and illnesses, insufficient food, lack of sanitation and personal hygiene items, and shortage of household supplies like fuel. Even with advance warnings, when a storm of this magnitude hits a nation with a densely populated coastal area like the Philippines, the damage, death, and injury toll is high. In all likelihood, there will be equally and more damaging typhoons in the future.

## THE FORMATION OF TROPICAL CYCLONES

Tropical cyclones in the North Atlantic, Caribbean, and eastern Pacific are known as hurricanes. In the Indian Ocean, the same type of severe storm is referred to as a monsoon, whereas in the western Pacific, China, Japan, and the Philippines, they are called typhoons. But these are all tropical cyclones, which are very large, warm, humid, rotating air masses. In order to be designated as a hurricane or typhoon, a tropical cyclone must generate wind speeds of at least 74 mph (119 km/h). Cyclones that do not develop these high winds are known as tropical storms and tropical depressions but can change their status without warning as they become fully developed. Some of the most damaging tropical cyclones have occurred in Southeast Asia, where population densities in exposed, low-lying coastal areas are often high and many of the buildings are of lightweight construction.

Despite centuries of disasters and decades of study, we do not fully understand how a tropical cyclone forms. Hurricanes, such as those that approach the Atlantic coast of the United States or pass through the Caribbean Sea, typically originate between about 5° and 25° N latitude (southern hemisphere tropical cyclones originate between 5° and 25° S latitude). In these tropical latitudes, the air is warm and air pressure gradients are weak. Many of these storms are born as hot air rises off the west coast of Africa and begins to move westward with the trade winds. Warm, moist air above the ocean flows into this air mass from below and is expelled at the top. As the warm air rises, moisture condenses, releasing both heat and rain while wind velocities increase.

In order for this column of air to rise high into the atmosphere, and for the characteristic rotating winds to form, the variation in wind velocities at different elevations must be small. This growing disturbance feeds on the warm tropical ocean and the hot tropical air, which slowly turns this swirling mass into a heat engine, spinning counterclockwise in the northern hemisphere (and clockwise in the southern hemisphere). As the warm air expands and cools aloft, it releases its heat. As this process intensifies, the storm can strengthen, although strong winds near the top of this column can shear off the top, causing dissipation or collapse of the potential storm. When one of these rotating spirals continues to develop and expand, it can evolve into a tropical storm and eventually a hurricane.

In the northern hemisphere, although these tropical cyclones rotate counterclockwise, overall they migrate or track in a clockwise direction, similar to the hemisphere's ocean currents. In the southern hemisphere, the patterns are exactly the opposite, with clockwise rotation but counterclockwise migration. Those tropical storms that form in the North Atlantic off West Africa continue to move westward, increasing their velocity and energy, and may develop into full-fledged hurricanes by the time they reach the Caribbean or the South Atlantic coast of the United States. Typhoons that form in the western Pacific will follow similar paths toward Asia (figure 3.2).

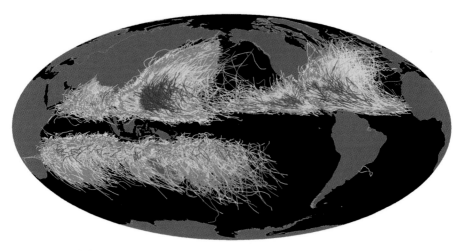

FIGURE 3.2. Global tracks of hurricanes over the past 150 years. Greens and blues are tropical depressions or tropical storms, and yellows, oranges, reds, and magenta are increasing intensities of hurricanes. (Courtesy of NASA Earth Observatory)

In any given year, ninety to one hundred tropical storms may form somewhere on the planet, but only five or six in the Atlantic or Gulf of Mexico will develop into fully fledged hurricanes and be given formal names. As summer transitions into autumn, residents of the Atlantic and Gulf Coast states often have unwelcome and poorly behaved guests with names like Andrew, Camille, Ike, or Katrina, who may tear their roofs off or, even worse, destroy their homes entirely. Hurricanes began receiving proper names in the 1950s, after an older system based on latitude and longitude was no longer seen as user-friendly. So in the United States the National Hurricane Center releases a group of names for each Atlantic hurricane season. This is a list of short, distinctive names selected to be easy to pronounce and culturally sensitive. After a particularly disastrous hurricane, Andrew in 1992 or Wilma in 2005, for example, a name can be retired so as not to be confused with subsequent events given the same name in later years.

### THE IMPACTS AND DAMAGE FROM HURRICANES

While relatively harmless at sea, unless you happen to be in a small boat far from shore, as these tropical cyclones approach land their true impacts and potential for devastation is quickly realized. They routinely cause billions of dollars in damage, can take thousands of lives, and have major impacts by modifying the shoreline through beach and dune erosion and deposition.

While the official Atlantic hurricane season extends from June 1 to November 30 each year, the peak of the season usually occurs in August and September when ocean water temperatures are the highest. Although we tend to think of the Gulf and South Atlantic coasts of the United States as the areas most exposed to hurricane damage in the United States, the paths of these tropical cyclones on occasion can reach as far north as New England, especially where areas like Long Island, Rhode Island, and part of Massachusetts extend into the Atlantic (figure 3.2).

About 61 million people live today in coastal counties that are vulnerable to hurricanes along the U.S. Gulf and South Atlantic coasts, and like most coastal regions the population continues to increase. From 1900 to 2015, there were 631 hurricanes that affected these counties as well as the Caribbean region, or 5.4 per year on average; 245 of these, or about two each year, have been classed as major hurricanes based on damage and death tolls. During this same 115-year period, there have been at least 92,630 fatalities (this is an underestimate, because death tolls for some older hurricanes are not known for certain), or an average of over 800 fatalities per year.

Hurricanes have been responsible for more loss of life in the United States than any other natural hazard. While those who live in the western portion of the United States, including Alaska, worry from time to time about earthquakes, large and damaging seismic events occur far less often than hurricanes, generally affect much smaller areas, and result in considerably fewer average annual deaths (about 20 per year). There has only been a single earthquake in the 240-year history of the United States that caused the death of more than 200 people, and this was the great 1906 San Francisco earthquake and fire whose fatality estimates are usually given as over 3,000. While virtually every state in the United States experiences flooding, the exposure and risk from individual flood events are considerably less than for hurricanes, and average annual deaths are far fewer, in fact the same average number of fatalities as from earthquakes (about 20 per year throughout U.S. history).

Overall damage due to hurricanes from 1900 to 2015 totaled nearly $490 billion, or an average of $4.2 billion annually. If hurricane costs are adjusted to 2005 for changing societal conditions, this number more than doubles, to about $10 billion per year. These losses are increasing in large part because more people are moving to hurricane-prone coastal counties but also because the investments in homes and other developments, and their values, are increasing. Six of the ten most destructive hurricanes between 1900 and 2006 occurred in 2004 and 2005, and then we can add 2012's Superstorm Sandy.

In 2010, the total population in counties along the Gulf and Atlantic coasts of the United States, extending from Texas to New York, that is exposed to hurricanes was 19 percent of the total U.S. population. These coastal populations are projected to increase about 10 percent, or an additional 6.1 million people, by 2020. The percentage increase in people sixty-five and older has been growing far faster (about two to four times faster) than the total population. Without some action to address the increasing concentrations of people and development in these coastal areas where hurricanes have historically made landfall, damage and loss of life will continue to increase.

## TYPES OF HURRICANE DAMAGE

Hurricane damage can be inflicted by high-velocity winds, by the impact of large storm waves, and by elevated sea levels and flooding, whether from storm surge (salt water) or from high rainfall (freshwater). Structures engineered to withstand hurricane force winds may well not have been designed to deal with flooding or submergence. In fact, most homes are not designed to withstand submergence. While Superstorm Sandy, because of some technicalities, was not classified as a hurricane, this made little difference when sea levels rose higher than ever recorded and flooded large sections of Manhattan.

### Wind

Wind velocities during hurricanes start at 74 mph (119 km/h), the lower threshold for classification as a hurricane, and increase from there. Velocities commonly reach 150 mph, and the maximum sustained wind speed ever recorded was during Super Typhoon Yolanda in the Philippines in 2013. The pressure exerted by the wind is proportional to the velocity squared, so that a doubling of the wind speed produces four times as much force. As a result, the destructive power of high-velocity winds can literally blow structures over or apart. Wind speeds of 150 mph exert pressure of about 116 pounds per square foot on a wall or roof. Wind generally produces much more damage than flooding, although once a roof is ripped off or windows are blown out, rain can enter a building and exacerbate the damage. There are many proven construction materials and methods that can help reduce wind damage: steep roofs without overhanging eaves; concrete, metal, or tile roofs, which are more resistant to wind than shingle roofs; laminated and tempered glass windows; and strong shutters or plywood covers over windows. Nonetheless, during severe hurricanes with very high velocity winds, all bets are off, and there are countless examples when entire oceanfront neighborhoods have been completely destroyed.

*Storm Surge*

The flooding of low-lying areas during hurricanes is usually caused by the interaction of several processes, including storm surge and large breaking waves. High wind velocity and low atmospheric pressure can allow the sea surface to rise near shore, in extreme cases as high as 25 feet above normal predicted sea level.

In the great Galveston Island hurricane of 1900 (known also as Isaac's Storm, named after the local weatherman who tried to warn the city), the storm surge reached nearly 15 feet above normal tides and washed completely over the island, which had an average elevation of about 9 feet above sea level. The city was completely unprepared and was nearly totally destroyed as large wooden houses and other structures literally floated away. The death toll was between 6,000 and 12,000 in a city of 36,000, making it the deadliest natural disaster in U.S. history. Hurricane damage adjusted to 2015 dollars has been calculated at $1.55 billion. In rebuilding Galveston, sand and mud were pumped out of the bay to raise the elevation of part of the island 17 feet above its former elevation, and a large seawall, 21 feet high, was built to provide protection for 10 miles of the eastern, more developed end of the 30-mile-long Galveston Island (figure 3.3). All protection usually ends somewhere, however, and the Galveston seawall does not protect the western end or the inland side of the island from storm surge.

One hundred eight years later, Hurricane Ike passed directly over Galveston Island. The storm surge reached up to 20 feet above normal tide, 5.2 feet above the 1900 hurricane level, and overtopped the seawall built after the devastating hurricane of 1900 (figure 3.4). Ike left behind $29.5 billion in losses, making it the third costliest Atlantic

FIGURE 3.3. The Galveston seawall was built to protect the city after the disastrous hurricane of 1900. (Photo: Gary Griggs © 2013)

hurricane at the time (figures 3.5 and 3.6). The backside of the island, which seemed to be protected but was at a significantly lower elevation than the area immediately behind the seawall, suffered major damage from storm surge.

This points out one of the engineering challenges of trying to protect infrastructure, communities, or cities from severe or infrequent natural hazards, whether hurricanes, floods, or earthquakes. No entity or agency can afford to provide complete protection from all potential natural hazards. So some design event, often the 100-year flood, is agreed on, calculations are made, and the protection is engineered or designed accordingly. Nonetheless, depending on the number of years of observations or measurements (e.g., the length of the flood record) or changing environmental conditions (e.g., rising sea level), the design event selected at one point in time may not be an accurate projection of the risks from the particular event in the future.

During Superstorm/Hurricane Sandy in 2012, the strength and angle of approach combined to produce a record storm surge in New York City. The water level recorded at Battery Park, at the southern tip of

FIGURE 3.4. High-water levels from major historic hurricanes are marked on the entry to a restaurant in downtown Galveston. (Photo: Gary Griggs © 2013)

Manhattan, topped 13.9 feet, exceeding the 10.2-foot record set by Hurricane Donna fifty-two years earlier. The East River (which is not really a river at all but a tidal channel) overflowed its banks, flooding large sections of Lower Manhattan. Seven subway tunnels under the East River were flooded, causing the greatest damage in the 108-year history of the Metropolitan Transportation Authority (figure 3.7). The storm surge flooded the Ground Zero construction site, and over 10 billion gallons of raw and partially treated sewage were released by the storm into the surrounding waters. Considerable damage to homes and other buildings, roadways, boardwalks, and mass transit facilities in low-lying coastal areas of both New York and New Jersey was caused by the storm surge, combined with large waves. Seventy-two people

FIGURE 3.5. Aerial view of shoreline damage in Gilchrist, Texas, from Hurricane Ike, January 2009, with only a single surviving home. (Photo: Courtesy NOAA / National Weather Service Houston Galveston, TX/Galveston County Office of Emergency Management)

died as a result of Hurricane Sandy, and total damage reached about $50 billion, making it the second costliest storm in U.S. history, after Katrina.

## Waves

Hurricane force winds blowing over large expanses of sea surface for several days can generate very large waves. During Hurricane Sandy, a buoy off New York measured a wave 32.5 feet in height, a record for that location, surpassing by over 6 feet the largest wave generated by Hurricane Irene, which passed through the same area in 2011. Large waves combined with a storm surge break closer to shore and intensify the damage that either might cause alone. Large hurricane-driven waves scour sand, eroding protective dunes and beaches and leaving only the structures themselves to withstand wave impact. Waves can damage or destroy homes, roads, bridges, and piers. Hurricanes can break through barrier islands, cutting new tidal channels and leaving some island areas completely cut off from the mainland. Building codes in

FIGURE 3.6. Bolivar Peninsula, Texas, before and after the passage of Hurricane Ike in September 2008 showing destruction of houses. (Photo: U.S. Geological Survey)

FIGURE 3.7. The flooded Brooklyn Battery Tunnel during Hurricane Sandy. (Photo: Timothy Krause licensed under CC BY 2.0 via Flickr)

some hurricane-prone areas require elevation on pilings or posts, but depending on the depth of embedment in the sand and the amount of beach scour, as well as storm surge and wind velocities, even this has not been enough to protect many structures. During Hurricane Sandy, entire beachfront communities built on sand in New York and New Jersey were flooded, tipped over, washed inland, or completely demolished by waves and storm surge (figure 3.8).

### Rainfall and Flooding

Many tend to think of hurricanes, cyclones, and typhoons as primarily coastal hazards, but the intense rainfall that occurs as these disturbances move inland can be even more damaging than coastal impacts. The size of the storm, the speed at which it advances, and the inland topography and vegetation or land cover exert the major influences on how much rain falls, over how large an area, how fast it runs off, and the downstream impacts. Where cyclones are large and slow moving and topography is steep, rainfall can be very heavy. Where mountains or topographic barriers exist near the coast, precipitation during a large cyclone can be extreme; many world rainfall records have resulted from these

FIGURE 3.8. A number of oceanfront houses were completely destroyed on Long Island, New York, during Hurricane Sandy. (Photo: Cheryl Hapke, U.S. Geological Survey)

conditions. The island of Réunion in the Indian Ocean is one example: a record 45 inches of rain have fallen in just twelve hours, 72 inches of rain have fallen in a single day, and 97 inches, or just over 8 feet, of precipitation fell in two days—all global records.

Tropical storms do not have to persist as hurricanes to produce significant flooding. In June 1972, Hurricane Agnes (which at the time was the costliest hurricane to hit the United States in recorded history) had deteriorated to a tropical storm by the time it passed over New York and Pennsylvania. The storm covered a circular area with a diameter of about 1,000 miles and generated the largest amount of rainfall the area had ever recorded. Severe flooding took place in Virginia, Maryland, Pennsylvania, and New York. Maximum precipitation of 19 inches in Pennsylvania forced a hundred thousand people to flee their homes, some buildings were submerged under 13 feet of water, and the governor's mansion was flooded. Fifty lives were lost and damage reached $13 billion (in 2015 dollars).

Tropical Storm Allison passed over Houston, Texas, in 2001 and dropped 37 inches of rain in thirty-six hours, which was three-fourths of the average annual rainfall for the city. Flooding in this fourth largest U.S. city damaged more than 45,000 homes and businesses. The storm left forty-one dead and over $12 billion in damage (2015 dollars) and became the costliest tropical storm in U.S. history.

## HURRICANE KATRINA

Hurricane Katrina, which hammered the Gulf Coast in late August 2005, was one of the deadliest hurricanes to ever strike a U.S. coast, causing the death of 1,833 people. Katrina also produced the greatest amount of damage, $108 billion, of any recorded storm. After passing over Florida, the storm wreaked havoc on Biloxi and Gulfport, Mississippi, then continued west across southeastern Louisiana with devastating storm surges ranging from 10 to 28 feet above sea level. The combined effects of large waves and the storm surge led to the collapse of the levees protecting New Orleans. Eighty percent of the city was ultimately flooded, displacing over a million people (figure 3.9).

One of the challenges of protecting New Orleans is that on average the city is about 6 feet *below* sea level. To make matters worse, because of its history of draining wetlands and extracting groundwater, as well as the overall subsidence of the Mississippi Delta from thousands of feet of accumulated sediment, the city is sinking and has been for decades. The tide gauge closest to New Orleans, at Grand Isle, has recorded a rise in sea level relative to land of 9.03 millimeters per year, or 3 feet per 100 years. This is about five times higher than the average global rise in sea level. To call this a serious problem would be an understatement, and it presents major challenges for this colorful and historic city, challenges that will only be amplified in the future as the level of the oceans continue to rise and the level of the land continues to sink.

New Orleans sits in a depressed bowl between the Mississippi River levees on one side and the levees around Lake Pontchartrain on the other (figure 3.10). Because of its low elevation relative to sea level and the reduction of much of the original natural protection that was provided by wetlands and barrier islands from storm surges, New Orleans is extremely vulnerable to hurricanes. The wetlands and barrier islands have been shrinking and eroding as a result of the channelization of the Mississippi River for shipping such that floods do not replenish the sediments as they did under natural conditions. Because of the city's location in this bowl, it takes a very long time to remove the floodwaters and begin to recover.

The impact of Hurricane Katrina on New Orleans was amplified by a combination of poor pre- and post-hurricane responses and poor

FIGURE 3.9. Flooding in New Orleans, Louisiana, from Hurricane Katrina, showing Interstate 10 looking toward Lake Pontchartrain. The breach in the 17th Street Canal that was responsible for much of the flooding is just off the left side of the photograph. (Photo: Petty Officer 2nd Class Kyle Niemi, U.S. Coast Guard via Wikimedia Commons)

rescue and recovery efforts at the local, state, and federal levels. A select bipartisan congressional committee charged with investigating the responses concluded that there was plenty of blame to go around. An aging and poorly maintained federal system of levees, failure of residents to respond to evacuation warnings, and slow and uncoordinated responses at every level of government all contributed to the disaster.

In response, levees were rebuilt higher and with a deeper support system, Congress passed legislation to reorganize the Federal Emergency Management Agency (FEMA) to make it more responsive, and New Orleans improved its emergency response and communication

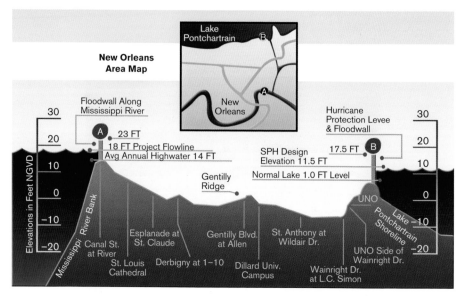

FIGURE 3.10. Topographic cross section of New Orleans, showing water level elevations on either side of the city. (Image: Alexdi at English Wikipedia licensed under CC BY-SA 3.0 via Wikimedia Commons)

system. Yet parts of New Orleans have never recovered. If we are to reduce losses from future events, we must learn from every natural disaster and ensure that the same mistakes, the same uncoordinated agency responses, and communication failures are remedied and do not recur.

## FUTURE HURRICANE VULNERABILITY OF THE ATLANTIC AND GULF COASTS OF THE UNITED STATES

As people continue to migrate to the Sunbelt states and home prices and property values continue to rise, exposure to and liability from future hurricanes has increased. The insured value of property along the Atlantic and Gulf coasts jumped from $7.2 trillion in 2004 to $10.6 trillion in 2012. Miami has some of the most expensive real estate in the nation, yet it is one of the five coastal cities most vulnerable to future hurricanes. And oceanfront construction is continuing at a record pace (see figure 1.9). Perhaps most ominous for Florida, projections are that populations in counties at or near the coast are expected to rise by 16 percent between 2010 and 2020, matched or surpassed only by South Carolina (17 percent), Virginia (18 percent), and Texas (19 percent), all hurricane vulnerable areas. A leading insurance spokesman reported at a National Hurricane Conference in 2015 that this projected population growth means not only more homes but also more businesses and more public buildings—schools and hospitals—and infrastructure in "harm's way" and that "too many people are in denial about the risk."

A key issue facing many East Coast cities, in particular, is that sea-level rise due to global warming and land subsidence is already making it more likely that damaging storm surge flooding will take place, even during weaker hurricanes. The following list of the top five most vulnerable U.S. cities to a hurricane landfall, and why, comes from a presentation from an insurer's perspective (insurance companies, for financial liability reasons, are usually on the front line of risk projection). The list takes into consideration storm frequency and historic tracks and city vulnerability, including the population living at or below storm surge elevations. This is certainly not a complete list and almost every city along the Gulf and East Coasts is a potential target for hurricanes. It does, however, include some of the potential hurricane hot spots that keep weather forecasters worrying.

1. Tampa–St. Petersburg, Florida

   The Tampa–St. Petersburg area, on the west coast of Florida, is extremely vulnerable to storm surge flooding and has been fortunate to escape a direct hit from a strong hurricane for many years. Climate Central calculates that the 100-year flood height in this area is 6.5 feet above the high tide line and that about 125,000 people are currently living below this flood level. In St. Petersburg alone, there are more than 45,000 homes that lie below 6 feet in elevation and would likely be vulnerable to a storm surge of that magnitude or greater.

2. Miami, Florida

   Miami has had its share of historic hurricanes, but residents have not experienced a major storm since Andrew in 1992, which flattened southern portions of Miami, including the communities of Homestead and Florida City. The combination of sea-level rise, population growth, and new construction in coastal areas is increasing the threat of storm surge–related coastal flooding. About 56,000 people in Miami Beach and 23,000 people in Miami live below Climate Central's Surging Seas 100-year flood height of 3 feet above high tide.

3. New Orleans, Louisiana

About half of New Orleans lies below sea level, and 340,000 people live below the 100-year flood level. Although Hurricane Katrina was a disaster for New Orleans because of levee and dike failure and because so much of the city is below sea level, a hurricane is likely to pass within 50 miles of the city every seven to eleven years on average. The ongoing loss of wetlands in southeastern Louisiana is continuing to remove the city's natural protection from future storm surges.

4. Norfolk/Virginia Beach, Virginia

While Norfolk and nearby Virginia Beach are often missed by the most intense storms, they are vulnerable to hurricanes that move up the Eastern Seaboard and were affected in 2011 by Hurricane Irene. The U.S. Navy's largest base is located on low-lying land in Norfolk. Due to subsidence of the land and increased rates of sea-level rise, the shoreline of Virginia's Chesapeake area is particularly vulnerable to flooding from future storm surges, with 40,000 people living below the 100-year storm level.

5. Houston-Galveston, Texas

Galveston was the site of America's worst recorded hurricane disaster, when Isaac struck in 1900, sending a storm surge completely across the island, killing between 6,000 and 12,000 people. Despite the continued risk of coastal flooding in Galveston, about 40,000 people live below the 100-year flood level. Hurricane Ike damaged Galveston and nearby Houston in 2008, causing $29.5 billion in damage. The communities of Crystal Beach, Gilchrest, and High Island (which apparently wasn't high enough) experienced near-total destruction of property.

LOOKING TO THE FUTURE: REDUCING HURRICANE LOSSES

Historically, hurricanes have been the costliest and deadliest natural disasters in the United States, surpassing earthquakes, tsunamis, floods, and anything else you can think of. Yet the coastal areas most vulnerable to hurricanes are typically regions that are both rapidly growing and popular tourism and vacation sites. This makes preparing for hurricanes even more challenging, yet we cannot put our heads in the sand and forget about the disasters of the past, nor can we simply hope that similar or more damaging events will not occur in the future. They will, although there is still some uncertainty about whether the climate change and global warming we are experiencing will increase hurricane frequency and intensity. Human activities, such as those producing greenhouse gas emissions, may have already caused changes that are not yet detectable because of the small amount of change, or because of our own observational limitations. However, anthropogenic warming by the end of this century will likely cause tropical cyclones globally to be more intense on average and to produce significantly higher rainfall rates than at present. A warmer ocean will provide more energy to the atmosphere and therefore raise the probability of more and larger hurricanes in the future.

## Natural Protection

In many undisturbed tropical or subtropical regions there are natural landforms and vegetation that help to reduce the impacts of storm surge and wave action. Mangroves, for example, which can live in salt water, are recognized for their ability to trap sediments and stabilize shorelines (figure 3.11). Mangrove forests can reduce risks from coastal hazards such as waves, storm surges and tsunamis. They have the ability to reduce both wave heights and flood depths and can lessen the damage landward of the forests. Mangroves occur worldwide in over 120 countries and territories, primarily between latitudes 25° N and 25° S, and most often along low-energy coastlines and within embayments and coastal lagoons. In the continental United States, mangroves are limited to the Florida peninsula and a few areas along the coasts of Louisiana and southern Texas. While not originally present in the Hawaiian Islands, they were introduced to stabilize mudflats on Molokai in 1902 and have subsequently invaded nearly all the other islands. While mangroves are now understood to play an important role as a buffer to the impacts of tropical cyclones, it is estimated that about 35 percent of mangroves across the planet have been removed or destroyed. Whether for shrimp or fish farming, wood products, or access to the shoreline for development, wholesale mangrove removal has devastated large areas of these unique trees, exposing the shoreline to the impacts of storm surges and waves. With recognition of the important role of mangroves, reforestation is beginning in some areas to rebuild these natural barriers. Thailand, Indonesia, and Florida are just a few areas where active mangrove restoration projects have been implemented.

FIGURE 3.11. Mangroves provide shoreline protection by trapping sediment and buffering the coast from wave attack and storm surge. (Photo: Carlos Andrés Reyes licensed under CC BY 2.0 via Flickr)

Dunes also can provide a partial barrier to storm surge and wave attack, although because they are just piles of unconsolidated sand, their effectiveness is limited, and they can be removed quickly under prolonged wave attack when sea levels are elevated. Coral reefs also provide a natural barrier in latitudes where they are healthy and continuous. In many locations, however, human activity or development has removed, altered, or otherwise reduced the extent or effectiveness of these natural barriers and buffers. Whether mangroves, dunes, marshes, or coral reefs, natural protection takes no maintenance, has no costs, and can repair or regrow after major tropical cyclone or storm damage, thus helping buffer shorelines at far less cost than virtually any engineering solution.

## Hurricane Warnings and Evacuation

Hurricanes, floods, and other weather-related hazards should no longer come as complete surprises, at least in the United States. We now have a robust system of orbiting satellites, specialized aircraft, and gauging stations for river levels and rainfall that allow us to detect, monitor, and track severe weather systems before they reach coastlines or populated areas. In contrast to earthquakes, where prediction has eluded us to date, our capabilities with regard to atmospheric phenomena are substantial. There is always uncertainty involved, however; tropical cyclones can change course, increase or decrease speed, and gather or lose strength along their tracks. Nonetheless, the U.S. National Hurricane Center attempts to provide 12- to 24-hour warnings of the probable paths of all tropical storms so that the precautions, evacuations, or other preparations can be initiated prior to landfall. This requires, however, that people listen or have access to the warnings and respond appropriately. Last-minute evacuations from barrier islands, which may be connected to the mainland only by a narrow bridge or occasional ferries, are risky. "Hurricane parties" of people who have never experienced a Category 4 or 5 hurricane and believe that they are immune from danger end very badly when there is literally nothing left of the party houses. Observing the magnitude of the damage in areas that have felt the direct impact of hurricanes should provide enough information and perspective on the wisdom of taking all precautions and leaving a potential impact area well in advance of predicted landfall.

## Construction Standards and Building Codes

There are no national building codes for dealing with hurricanes, so it falls to state or local government agencies to develop and then enforce their own standards and codes. With any large natural disaster accompanied by major property losses, there is often an immediate legislative reaction aimed at developing more stringent building standards. There is typically a narrow window, however, in which to accomplish these changes, and then another issue rises on the legislative priority list and the disaster becomes a distant memory. When proposals are made to strengthen building codes, it may be years before they are approved or implemented, and there will nearly always be opposition from developers, speculators, or builders who are not worrying about living in the structures they are building and selling. History has shown that where stringent standards have been adopted and enforced, damage can be held to a minimum under most storm conditions. Florida's Coastal Construction Control Line is a good example. This Control Line delineates shoreline areas that are subject to 100-year storm impacts, whether

erosion, flooding, or other forces, and requires that any habitable structure meet certain construction standards. Experience has shown that
homes can be significantly strengthened in order to better resist the
impacts of hurricane forces at only modest increases in construction
costs. Nonetheless, when a shoreline community or typical home takes
a direct hit from a Category 5 hurricane such as Ike, the probability of
suffering major damage is still very high. Very few typical wood frame
or lightweight structures are a match for 180 mph winds and storm
surges of 6 or 8 feet. Building or buying directly on the back beach or
low frontal dunes carries some very real risks that need to be understood. The sand beneath your foundation was probably deposited there
by waves or wind in the not too distant past, and the likelihood of
waves and storm surge impacting the area in the future is very high, and
increasing with each additional inch of sea-level rise.

CHAPTER 4

# Storms, Waves, Coastal Erosion, and Shoreline Retreat

Anyone who has watched winter storm waves batter a shoreline can appreciate the tremendous power exerted by the ocean. Regardless of which ocean or which hemisphere, the waves along with daily tidal fluctuations and longer-term sea-level rise are combining their efforts to impact coastlines globally. Whether it is the massive Pacific, large enough to hold all of the continents with some room left over, the Atlantic and Indian Oceans, or even smaller seas like the Mediterranean, Caribbean, or North Sea, storm waves frequently reach the shoreline with enough power to produce major damage and widespread destruction.

The oceans of the world wash over the shorelines of approximately 114 nations on six continents, as well as a large number of low-lying island nations or territories. From New Zealand to New Jersey, Costa Rica to Cambodia, and Singapore to Sri Lanka, the ocean provides important benefits but also can generate negative impacts. Marine waters provide nutrients for marine life and support important fishing industries, provide easy access for trade and commerce, and in many places are at the core of important tourist and recreational economies. Yet these same coastal waters are not always calm and passive, and with winter storms, elevated sea levels, and the not infrequent El Niño, hurricane, cyclone, typhoon, or tsunami come threats to all the public infrastructure and private development that civilization has built along shorelines everywhere.

FIGURE 4.1. High natural dunes were overwashed during Hurricane Sandy at Fire Island, New York, carrying large amounts of sand to the island's interior. (Photo: Cheryl Hapke, U.S. Geological Survey)

## SANDY SHORELINES: BEACH EROSION AND SHORELINE RETREAT

Coastal erosion means different things depending on where you live or have traveled. For the residents of the Gulf Coast or the Mid- to South Atlantic coasts of the United States, where the shoreline consists of a nearly flat coastal plain, typically fronted by offshore, low-relief, sandy barrier islands, coastal erosion occurs in the form of barrier island washover and landward migration. Because the barrier islands consist only of loose sand, erosion or migration can take place rapidly, particularly during hurricanes or nor'easters, when islands can be completely overwashed, new inlet channels can be scoured and old ones filled in (figure 4.1).

Change can happen quickly during extreme events on barrier island shorelines, as well as from the day-to-day forces of the waves on the loose sand. Depending on whether the island is undeveloped, like 36-mile-long Assateague Island and National Seashore just off the coast of Maryland, or intensively developed, like Ocean City on adjacent Fenwick Island to the north, severe events and even the less energetic summer waves can move a lot of sand around and modify the island shorelines over time. Habitats, whether marsh, dune, coral reef, or mangrove forest, as well as the animals that inhabit these environments, have adapted and migrated throughout history. They have had to, and hopefully most of them will adjust and survive with future changes. It is the fixed human development that is in trouble. Cities do not migrate easily or quickly.

Migration or shoreline change rates can be very high along these sandy barrier islands, simply because sand does not provide the resistance to wave attack that the hard rock coastlines of places like Maine do. Beaches, whether in Florida or France, California or Costa Rica, can undergo several different types of "erosion." Most common is the seasonal or temporary summer to winter erosion, followed by accretion the next spring and summer (figure 4.2). Driven by seasonal differences in wave energy, these beach cycles are expected and should not come as a surprise, unless people are recent arrivals to the coast, and then the loss of the beach in front of their new homes may in fact be a shock. These seasonal fluctuations may involve the narrowing of the beach by 100 feet or more and the lowering of the sand levels by 6 to 8 feet. During the next spring and summer, with the return of lower and less energetic waves, the sand will migrate back onshore and the beaches will expand. All summer beaches will not be the same width, nor will the winter losses be uniform from year to year, but the general patterns are expected and quite predictable.

Beaches can also undergo more permanent erosion or narrowing where sand supply is either temporarily or permanently reduced. This process can be regional in scope where the delivery of sand to the shoreline has been eliminated by the construction of dams or inland debris basins on rivers or streams. There is a long history of dam construction in the United States. Today in China and Africa more and more dams are being planned, typically without serious consideration of their often-negative impacts. One of those impacts is a significant loss of the sand that has historically been transported to the coastline to nourish beaches. Water supply, hydroelectric power generation, recreation, flood control, and even navigation have nearly always outweighed any downstream effects of dams. With sand from rivers and streams providing a large proportion of the planet's beach sand, the losses can be significant and have far-reaching effects. Sand mining for construction purposes and dams on the Magra River in the Lucca province of Italy led to as much as 300 feet of shoreline erosion between 1938 and 1998. The intensively used beaches as far as 4 miles downdrift from the river

FIGURE 4.2. Typical seasonal differences in beach profile: (a) summer and (b) winter. (Photos: Gary Griggs © 2016)

mouth were narrowed, and virtually this entire shoreline has now been armored with hard structures (figure 4.3).

China is home to the greatest number of dams, 20 percent of the world's large dams (over 50 feet in height), with 45 percent of these used for irrigation. Nationwide in the United States, there are nearly 80,000 dams according to the Army Corps of Engineers National

FIGURE 4.3. Forti dei Marmi on the Tuscany coast of Italy has a sand deficit. A combination of groins and submerged breakwaters between groins has been constructed in an effort to retain the beaches. (Imagery © 2015 Google, TerraMetrics, Map data © 2015 Google)

Inventory of Dams. This amounts to a dam built almost every day since the Declaration of Independence was signed in 1776. While these dams vary widely in size, almost all of them trap sand destined for the shoreline. In California, more than 500 dams impound more 16,000 square miles (38 percent) of the state's coastal watershed area. These reservoirs have now collected about 200 million cubic yards of sand, enough to build a beach 100 feet wide, 10 feet deep, and 1,045 miles long, virtually the entire length of the state's coastline. The average annual delivery of sand and gravel has been reduced by about 3.7 million cubic yards per year, or 25 percent, much of this along the more urbanized Southern California coast where the state's beach recreation and tourism are concentrated.

Sand impoundment behind large coastal engineering structures has also been well documented globally, particularly along the shoreline of the Mediterranean but also along the East and West Coasts of the United States. Much of the shoreline erosion and migration along the Atlantic barrier islands has been ascribed to long jetties that have been constructed to stabilize channel entrances between islands for navigation. These include the jetties at Ocean City, Maryland, between Fenwick and Assateague Islands. Port Everglades–Fort Lauderdale, Boca Raton Inlet, Boynton Beach Inlet, Palm Beach Inlet, Jupiter Inlet, and Fort Pierce Inlet are all good examples along the east coast of Florida where littoral drift (the transport of sand along the shoreline from wave action) from one direction (usually the north) has accumulated against the updrift jetty, causing the beach across the inlet on the downdrift side to narrow. These shoreline responses are expected and predictable now that we have decades of beach accretion and beach erosion observations at these inlets. Along almost any shoreline where long jetties are proposed, littoral drift will be trapped, and downcoast retreat or erosion can usually be expected. These impacts need to be evaluated for their probable long-term impacts before any new major coastal barriers are contemplated and planned. Unfortunately, short-term economic or political interests have often prevailed over long-term experience and observations.

Littoral sand transport along the coast of California is interrupted frequently, particularly along the more populated and developed Southern California coastline. A number of large breakwaters and jetties have been constructed over the past eighty-five years to develop and protect small craft harbors and marinas. With the exception of a few (Moss Landing, Redondo King, and Dana Point Harbors), which were built near the heads of submarine canyons where littoral sand is drained offshore to the deep-sea floor, most of these harbors instantly become traps for littoral drift. Sand has accumulated against either a breakwater (Santa Barbara, Oceanside; figure 4.4) or the updrift jetty (Santa Cruz, Port Hueneme; figure 4.5) or against a jetty and behind a breakwater (Ventura, Channel Islands, and Marina del Rey), with at least the short-term loss of downcoast beaches until regular and expensive harbor dredging was initiated and sand was returned to the littoral system.

The influx of large numbers of people to coastal states, particularly locations like Florida and California, with many of them wanting pleasure boats, has led to the construction of many new marinas and the expansion of existing harbors over the past fifty or so years. The effects on littoral drift and the potential for downcoast beach erosion were not often a significant issue decades ago when local interests lobbied for small craft harbors. Today we have the experience, the observations, the aerial photographic record, and the knowledge of littoral drift direction and approximate rates in many locations so that with some preliminary research we can get a reasonably accurate idea of what to expect in any

FIGURE 4.4. Millions of cubic yards of sand accumulated against the breakwater at Santa Barbara, California, following harbor construction in 1930. The football stadium, parking lots, and all buildings to the left of the harbor were built on what was formerly seafloor. (Photo: Courtesy of Bruce Perry, Geology Department, California State University Long Beach)

FIGURE 4.5. Following construction of the jetties at the Santa Cruz, California, small craft harbor in 1963–65, littoral sand was impounded west (to the left) of the jetties and has widened the beach for nearly a mile upcoast. (Photo: Joel Avila © 2006, Hawkeye Photography)

particular location where jetties, a breakwater, or other littoral obstructions are proposed. Littoral drift or alongshore transport rates of sand have often been estimated or determined from harbor dredging rates, or the rate of accumulation of sand behind a littoral barrier such as a groin, jetty, or breakwater.

There is no excuse or rationale to naively undertake a large shoreline engineering project, anywhere in the world, without careful evaluation of the particular coastal environment and consideration of the impacts of littoral transport and beach erosion. Major obstructions placed across the beach, whether breakwaters, jetties, or groins, are almost certain to trap littoral drift and alter the shoreline and beaches. If we are to avoid the downcoast shoreline impacts and the high costs of the typical annual dredging of channel entrances or inlets, these potential impacts and their associated costs need to be understood at the front end. These should no longer be unknowns or surprises and need to be factored into operating costs.

There is one additional type of shoreline erosion or retreat along sandy barrier island coasts, that of complete migration of an island. These islands began life about 20,000 years ago as the last ice age drew to a close. At that time, sea level globally was about 350 to 400 feet lower than today and the shoreline was located out at the edge of the continental shelf. There were beaches, dunes, and other features just like we see today along the South Atlantic and Gulf Coasts, but they were dozens of miles offshore from today's shoreline. As sea level gradually rose, these sand accumulations migrated landward along with the shoreline, rising above the surface of the ocean in places as sand bars, sand spits, and low dune-covered islands. Sea level stabilized about 7,000 to 8,000 years ago, and waves, wind, and storms grew these barrier islands, building many of the features we see today. Examples are Galveston Island and Padre Island along the Texas coast; Canaveral National Seashore, Miami Beach and Palm Beach, Florida; Georgia's Sapelo Island and South Carolina's Daufuskie Island; the Outer Banks of North Carolina; Cobb and Hogg Islands off the Virginia shore; Ocean City, Maryland; Atlantic City, New Jersey; and Fire and Coney Islands of New York.

There are about 282 barrier islands along the U.S. Gulf and Atlantic coasts, stretching from Texas to New York. They are typically one to 3 miles wide, 6 to 60 miles long, several to 20 miles offshore, and often only 10 to 20 feet above sea level. These sandy islands are highly mobile and migrate landward over time as they are overwashed by hurricanes and nor'easters.

FIGURE 4.6. The Cape Hatteras Lighthouse prior to being moved inland 2,900 feet in 1999 due to shoreline erosion. (Photo: Mike Booher/National Park Service)

Cape Hatteras Lighthouse on the Outer Banks of North Carolina is a good example of the well-documented migration of a barrier island. When the lighthouse was built in 1870, it was about 1,500 feet inland from the shoreline or outer edge of the island, thought to be a safe distance at the time. The structure is an iconic lighthouse with its barber pole striping. At twenty-one stories, it is the nation's tallest lighthouse and the tallest brick lighthouse in the world. By 1919, fifty years after it was built, the shoreline had migrated landward or inland over 1,100 feet. The island migrated another 200 feet by 1935, leaving the famed lighthouse just 115 feet from the shoreline. In that sixty-five-year period, the shoreline moved inland or westward 1,375 feet, or an average rate of 21 feet per year.

As the shoreline began to approach the base of the lighthouse, several different efforts were implemented that were intended to trap littoral drift and widen the beach. Groins were built, sand was imported, and artificial seaweed was placed offshore, none of which had much permanent effect on widening the beach. After several other proposals, and lots of debate, the National Park Service finally decided to pick up the lighthouse, all 4,800 tons of it, and in 1999 moved it 2,900 feet inland along a runway and track (figure 4.6). The $17.5 million effort (in 2015

FIGURE 4.7. Construction of jetties at Ocean City, Maryland, in the 1930s trapped the southerly littoral drift, widening the upcoast beach, cutting off sand supply downcoast and leading to 1,600 feet of shoreline retreat. (Photo: NASA)

dollars) has provided a safe home for the immediate future, with the hope that it will survive for another century, which is entirely dependent on future sea-level rise and hurricane and storm climate.

Jetties were built in the 1930s to stabilize the Ocean City inlet between Fenwick Island and Assateague Island off the coast of Maryland. In the early 1900s the two islands were actually joined, but a severe hurricane in 1933 created a new inlet and two islands. While littoral transport would have filled the inlet over time, a decision was made to stabilize the gap to provide better access and navigation to Ocean City. The jetties that were constructed trapped the southerly flowing littoral drift such that Fenwick Island to the north accreted while Assateague Island to the south had its sand supply cut off. By 1980 Assateague had retreated or migrated about 1,600 feet, or nearly 34 feet a year on average (figure 4.7). The island has a mean elevation of just 6 feet above sea level so is highly susceptible to storm overwash.

## ROCKY COASTS: CLIFF EROSION AND COASTAL RETREAT

Coastal erosion or retreat is different from beach or shoreline erosion and is defined here as the actual landward retreat of a coastal cliff or bluff. It is distinct from beach erosion in that it is not recoverable, at least within our lifetimes or by any natural processes. The words *cliff* and *bluff* are often used interchangeably, but for this discussion and future use, *cliff* refers to coastal landforms that consist of harder and more resistant rocks that stand higher and steeper than coastal bluffs, which are generally composed of weaker materials that are more prone to erosion and failure and are usually lower in relief and stand at gentler slopes. Cliff or bluff failure is a major coastal hazard around the world and threatens any buildings, improvements, or infrastructure located on this retreating edge.

The rate at which a cliff or bluff erodes depends on several different factors, which can be thought of as either *intrinsic,* or internal, which includes all the various physical properties of the geologic materials making up the cliff; and *extrinsic,* or external, which includes all the forces or processes affecting that particular stretch of coastline. For a rock cliff, the intrinsic or physical properties include the hardness or degree of consolidation or cementation of the cliff rock and the presence of internal weaknesses such as fractures, joints, or faults, as well as groundwater seepage, all of which can directly affect the resistance of the rock to wave impact or other forces. External factors include the amount of wave energy reaching the cliff, tidal range and sea-level rise, and rainfall and runoff, to name some of the most important processes.

Cliff erosion takes place through several different processes, gradually or sometimes instantaneously, taking their toll on private or public oceanfront property and any buildings or infrastructure present (figure 4.8). Hydraulic impact is perhaps the most important process and is simply the direct force of a breaking wave on the cliff. Where rocks are unconsolidated, weathered, jointed, or otherwise weak, waves can dislodge large and small fragments or blocks, leading to gradual failure or retreat of the cliff. Gravel and cobbles can be hurled against the cliffs under large wave conditions, further contributing to cliff breakdown and retreat. The sand, gravel, and even cobbles that the waves wash back and forth across the shoreline also become important abrasion tools. The constant grinding of the rocks against the bedrock works just like sandpaper but on a very large scale.

FIGURE 4.8. Despite the construction of concrete seawalls on the beach to protect the homes and support septic tanks, storm waves have overtopped and damaged the seawalls, eroded the bluff, and led to the loss of most of the houses at Gleason Beach, Sonoma County, California. (Photo: Kenneth and Gabrielle Adelman © 2009, California Coastal Records Project, www.Californiacoastline.org)

The bedrock exposed in cliffs and along the shoreline is broken down by a combination of physical disaggregation and chemical reactions. The alternating daily and seasonal cycles of wetting and drying and heating and cooling that take place within rocks in the intertidal zone, as well as the chemical breakdown of the rocks by seawater, are also important processes playing a role in cliff degradation and retreat. All these processes act in concert to either weaken the cliff-forming materials or break off and remove individual fragments.

Even during calm weather, small waves constantly wash sand, gravel, or shells across the intertidal zone—wetting, drying, and gradually weakening the rocks that make up the base of the sea cliff and carrying off the bits and pieces that break loose. Although this day-to-day activity takes its toll on the cliffs or bluffs, it is generally the winter storm waves at times of high tides or elevated sea levels that lead to major episodes of coastal retreat. In part, this is because the winter waves are larger and have more energy. In addition, prolonged winter rains weaken or can saturate the bluffs or cliffs, often making them more susceptible to failure. Another important factor is the reduction in width or even the total

FIGURE 4.9. Cliff erosion undermined these apartments in Pacifica, south of San Francisco, which were subsequently demolished. (Photo: Gary Griggs © 2013).

disappearance of protective beaches during the winter. With this buffer zone of sand reduced or gone, the waves can attack the cliffs, bluffs, or dunes more frequently and with greater energy (figure 4.9).

During severe winters, such as the El Niño conditions that periodically affect the West Coast of the United States, the nor'easters that attack the coastline of New England, or the storms in the North Sea that have been ravaging the Yorkshire coast of Britain for centuries, the evening news, newspapers, and now social media are packed with stories of houses or roads being undermined or collapsing into the ocean. Cliff or shoreline erosion is a logical consequence of wave attack at high tide, which is acting in concert with an accelerated rise in sea level, and should no longer be seen as an act of God or a surprising occurrence. History can tell us a lot if we take the time to read, study, and understand it.

Determining the long-term rate of cliff or bluff retreat is an important initial step in evaluating the long-term stability of any parcel of coastal land for development or construction. By using accurate historical parcel or survey maps, vertical aerial photographs, or, now, LIDAR (Light Direction and Range) surveys, which are becoming more common, an experienced geologist or coastal engineer can determine the change in the position of the edge of the coastal bluff or cliff over time. The processes that lead to cliff or bluff failure are episodic in time, however. Documented or measured retreat rates of a few inches to a few feet per year in sedimentary rocks are common but need to be seen as long-term averages and not an annual prediction.

## SEA-LEVEL RISE: COASTAL EROSION AND SHORELINE RETREAT

Eighteen thousand years ago, at the end of the last ice age, coastlines and shorelines around the world were several to over a hundred miles farther seaward. At that time, over 10 million cubic miles of seawater had been evaporated from the ocean and had accumulated on the continents as ice sheets and glaciers. This lowered sea level globally about 350 to 400 feet, moving the shoreline out to the edge of the continental shelf. As the last ice age drew to a close and climate began to warm, glaciers and ice sheets melted and retreated and seawater expanded as it warmed. The increasing ocean volume began to move the shoreline landward across the continental shelf. Much of this global coastal retreat took place between about 18,000 and 8,000 years ago, and then climate change slowed and the rate of sea-level rise leveled off. Based on the geologic evidence that has been collected, there was very little change in sea level over the past 8,000 years, which corresponds to virtually the entire history of human civilization. Deltas, coastal plains and their fertile soils, and access to the sea provided stability and allowed agriculture to develop as early humans began to transition from hunter-gatherers to farmers.

However, with the coming of the industrial revolution and the internal combustion engine in the nineteenth century, accompanied by the increased burning of fossil fuels, greenhouse gas emissions began to rise and global temperature increased, with sea-level rise following a short distance behind. Whether beaches or barrier islands, cliffs or coral reefs, coasts around the world are increasingly feeling the effects of sea-level rise and are experiencing retreat and erosion, which is explored at greater length in chapter 5.

Along cliff- or bluff-backed coasts, a continuing rise in sea level translates into high tides and waves reaching the base of the cliff more often, leading in most places to gradually accelerating erosion and retreat. Where the shoreline consists of low-relief beaches, tidal marshes

333333333

and estuaries, barriers islands, and coastal dunes, a rising sea means more frequent flooding, followed by permanent inundation over time and inland migration of the shoreline.

## RESPONDING TO COASTAL EROSION OR SHORELINE RETREAT

With the thousands of miles of developed beaches, dunes, barrier islands, and bluffs around the world, there are countless locations in virtually every coastal nation on the planet where development of all sorts, private and public, new and old, where erosion and retreat either are already threatening or will in the decades ahead. Future losses will be very, very high, and many now believe that the threat to coastal cities and low-lying development around the world from future sea-level rise, combined with storms, erosion, and inundation, will be the biggest threat that human civilization has ever faced.

Future sea-level rise, however, is a longer-term issue, at least for now. Throughout the twentieth century, those responsible for developing coastlines around the world responded to the hazards of coastal erosion or shoreline retreat in several ways:

1. *Do nothing or wait and see*
2. *Armoring or hardening of the shoreline*
3. *Beach nourishment*
4. *Managed or unmanaged retreat*

Each of these has advantages or disadvantages, and different geographic areas, political entities, cities, states, or nations have either intentionally or unintentionally made decisions to use one or several approaches. The "do nothing" or "wait and see" approach clearly has the lowest cost but also the greatest risk or potential consequences. We cannot predict when any particular structure built on the back beach or on the edge of an eroding bluff will be finally threatened, but doing nothing almost guarantees that the day will come when it is too late and damage or complete loss will result (see figures 4.8 and 4.9).

Historically, armoring the shoreline has been the most common approach to coastal erosion or shoreline retreat. By 2000, 10 percent of California's 1,100-mile-long coastline had been armored, although for the more urbanized shoreline of Southern California, this value climbs to 33 percent (figure 4.10). Whether rock revetments or seawalls or any

FIGURE 4.10. A riprap revetment was emplaced at Broad Beach in Malibu, California, but was considered temporary protection from shoreline retreat until a permanent solution could be planned and approved. (Photo: Gary Griggs © 2012)

of a variety of other engineered or non-engineered structures, there is a long history of coastal armoring, but in many cases these were not built with any or many of their potential effects in mind. These effects include visual impacts; loss of public beach due to placement of the structure on the sand; loss of the sand previously provided by the eroding cliff, bluff, or dune being armored; loss of public access to or along the beach; and passive erosion, or the gradual loss of the beach fronting the armor with a continuing rise in sea level.

In the United States some states have banned hard structures altogether, while others have made it more difficult to get a permit unless the primary structure (e.g., a house) is under imminent threat. The era of routine armoring of any eroding stretch of coastline in the United States is ending, as the negative impacts of protective structures have been recognized and understood and the inevitability of future sea-level rise becomes more obvious. While armor can provide short- or intermediate-term protection for private property and public infrastructure, with a changing climate and a rising sea there are no future guarantees that today's armor will survive far into the future. Armor, whether a rock revetment or a concrete seawall, has been emplaced to protect individual homes, small groups of homes, or even entire developments. A 50-foot-wide bluff-top parcel can be protected, or the same rock or concrete seawall can be a mile long and armor dozens of homes.

Another approach used for temporarily forestalling shoreline retreat or beach erosion is to artificially widen the beach with sand from some outside source. Beach nourishment, however, is almost always carried out as a very large-scale project, where thousands of feet of shoreline are at least temporarily widened with hundreds of thousands of cubic yards of sand. Beach nourishment has been employed for decades along the low-relief, typically barrier island–backed sandy shorelines of the Atlantic coast of the United States (figure 4.11). The largest recipients have been Florida, New York, and New Jersey, which have received a total of 315 million cubic yards of sand, much of this funded by the federal government. Adding in the remaining Atlantic coast states, the amount of sand dredged from offshore and dumped on the beaches totals about 430 million cubic yards. This is a difficult volume of sand to visualize; it is enough sand to build a beach 150 feet wide, 10 feet deep, and 1,460 miles long, or a beach extending all the way down the Atlantic seaboard from Rhode Island to the southern tip of Florida. Much of this sand has been moved around at federal expense and in recent years, with federal budgets being stressed, these projects have been more difficult to fund.

While sand has been placed on many beaches in California, with a few exceptions this has usually been labeled opportunistic and has been carried out as by-products of the construction of harbors and marinas, the maintenance dredging of harbor entrances and river channels, or other coastal construction projects. The beaches of Southern California benefited from about 130 million cubic yards of sand as by-products of various coastal engineering or maintenance projects between 1930 and 1993. Two nourishment projects in San Diego County (2001 and 2012) were carried out with the specific purpose of beach widening, for both coastal protection and recreation. Although $46 million were spent

FIGURE 4.11. A beach nourishment project under way at Nag's Head, North Carolina, where 4.6 million cubic yards of sand dredged from offshore were added to 10 miles of shoreline. (Photo: Courtesy of Coastal Science and Engineering, Inc.)

dredging 3.6 million cubic yards of sand from offshore and placing it on county beaches, within a year or so nearly all of the sand had been removed from the exposed beach by wave action.

Whether in New Jersey, Florida, or California, while billions of dollars have been spent moving sand from offshore to the shoreline for both recreational and shoreline protection benefits, the life span of this artificially added sand has in many cases been relatively short, in some instances less than a year. Knowing typical littoral drift rates along a specific stretch of shoreline can often provide a reasonable estimate of how long a known volume of nourished sand may remain. Beach nourishment should not be seen as a permanent or even long-term solution to beach or bluff erosion but simply a means of buying a little more time at great public expense.

## LOOKING TO THE FUTURE: PLANNED OR MANAGED RETREAT

In recent years, especially with the added impact of an increase in the rate of sea-level rise on already retreating shorelines, as well as several natural disasters on the Atlantic and Gulf Coasts (Hurricanes Sandy and Katrina) and in Japan (the Tohoku earthquake and tsunami) and several decades of large El Niño events along the California coast, damage and destruction to coastal homes and other structures has become quite commonplace. With sea level expected to rise between 18 and 52 inches by 2100 and to continue for centuries into the future, virtually every shoreline development, community, and city in the world needs to begin to think in the long term. Where structures are located in vulnerable coastal locations and have experienced or are threatened by flooding or cliff retreat, relocating or removing the structure is going to become a more important consideration or response. There are many existing developments where there are no other reasonable or acceptable alternatives, where future damage or destruction is almost guaranteed, and where rebuilding or protection is simply not possible or cost-effective. Certainly the size, condition, and physical setting of the dwelling or building are important considerations and need to be assessed by professional building movers and possibly a structural engineer. There are many examples of relocation or planned retreat, whether homes or other structures. The Cape Hatteras Lighthouse in North Carolina is a good example of what can be done for $17.5 million.

The highly erodible Yorkshire coast of England, where the bluffs consist of very weak glacial sediments, has waged a long battle with the storm waves of the North Sea and provides a useful and long-term perspective. Records for the location of the coastline and coastal settlements in the United Kingdom go back almost two thousand years, to the time of Roman occupation. There is an extraordinary record of Yorkshire's struggle against the waves, documented in a book published over a century ago, *The Lost Towns of the Yorkshire Coast* (1912), in which the author shows the progressive disappearance of twenty-eight towns during the previous two thousand years (figure 4.12).

Average annual retreat rates of the weak bluffs along the Yorkshire coast have been documented at about 6 feet per year, with rates in recent years being significantly higher (figure 4.13). The English have taken a new approach over the past decade or so, however, where caravan and mobile home parks have replaced the former villages and also the cow and sheep pastures (figure 4.14). Many city residents have purchased mobile homes and then leased concrete pads with utility connections in one of the dozens of caravan parks. These have become their summer and vacation home sites. Each of these parks is like a small city, often with grocery stores, pubs, casinos, and movie theaters so that the vacationers or even permanent residents have little reason to leave. And

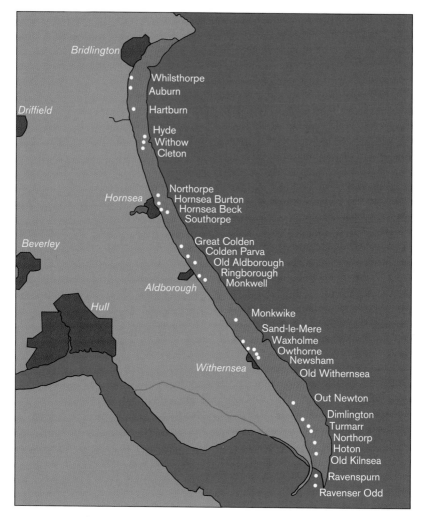

FIGURE 4.12. Thirty-one towns along the Yorkshire coast of the United Kingdom have been lost over the past 2,000 years (area in brown). (Map courtesy of and modified from Caitlin Green)

when the bluff edge gets close to the outer row of mobile homes, they simply hook up a truck and tow them a few hundred feet inland to a new, safer location (figure 4.15). This seems to have become an acceptable and resilient solution along the eroding coastline of Yorkshire, and to all appearances, it seems to have worked out quite well.

Nearly all coastlines are facing the problems of coastal and shoreline retreat, whether from the long-term but accelerating rise in sea level or

FIGURE 4.13. The erodible bluffs along the Yorkshire coast have been retreating at average rates of about 6 feet per year for the past 2,000 years. (Photo: Gary Griggs © 2015)

FIGURE 4.14. The caravans, or mobile homes, along the Yorkshire coast are simply moved back from the edge to a new location farther inland as erosion threatens. (Photo: Gary Griggs © 2015)

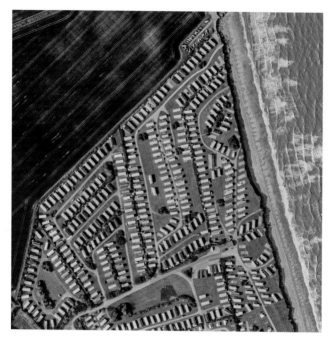

FIGURE 4.15. The empty concrete slabs at the bluff edge along the Yorkshire coast are where caravans have been moved farther inland. (Imagery © Infoterra Ltd & Bluesky, Getmapping plc, Map data © 2016 Google)

the shorter-term hazards of severe storms with large waves at times of high tides. While there are a few locations where shorelines are actually accreting (deltas or active volcanic islands like Hawai'i, which have their own problems) or where the coastline is being uplifted faster than sea level is rising (such as many northern latitude areas, which are still experiencing glacial rebound from the last ice age), these are unique, and most other areas are going to have to deal with this challenge for centuries to come. The options are limited, but the solution that will eventually be forced on us is managed retreat (see chapter 5).

We need to determine which areas are the most vulnerable to coastline erosion or shoreline retreat, eliminate federal subsidies for repeated reconstruction in harm's way, and then plan for an orderly process of moving structures and infrastructure back from the edge. We cannot continue to fight the inevitable rise in sea level. There are certainly structures, facilities, or infrastructure, large international airports, for example, many of which were built very close to sea level on artificial fill, that we may try to protect as long as possible. But a time will come, at some uncertain date in the future, when higher and higher walls will no longer work and we will be forced to move back to accommodate a rising ocean. It's not a question of if but when.

# CHAPTER 5

# Climate Change and Sea-Level Rise

Every nation on Earth with a coastline should be worried about sea-level rise. Throughout virtually the entire history of civilization, the past eight thousand years, more or less, depending on how we define civilization, the level of the ocean, and therefore the location of the shoreline, did not change much. While the tides did go in and out every day and storms and hurricanes temporarily elevated sea levels regionally, the overall level of the ocean stayed pretty much the same.

This started to change in a significant way about 150 years ago as humans began to use fossil fuels in increasingly larger quantities. Burning coal, oil, and natural gas, as well as wood, peat, and animal dung, produces carbon dioxide. With 85 percent of our global energy being produced from burning fossils fuels, the generation of carbon dioxide ($CO_2$) is still increasing. In 2014 our global population emitted about 40 billion tons of $CO_2$, twice as much as in 1980 and ten times as much as in 1930. Forty-four percent of this came from just two countries, China and the United States. Of that yearly production, China is now the global leader, with 28 percent of the total; the United States is second with 16 percent; and the European Union generates just 10 percent. India and the Russian Federation each produce 6 percent; Japan, 4 percent; and all the other nations, the remaining 30 percent (figure 5.1). From 2005 to 2014 about 44 percent of that carbon dioxide accumulated in the atmosphere, 26 percent in the ocean, and 30 percent on

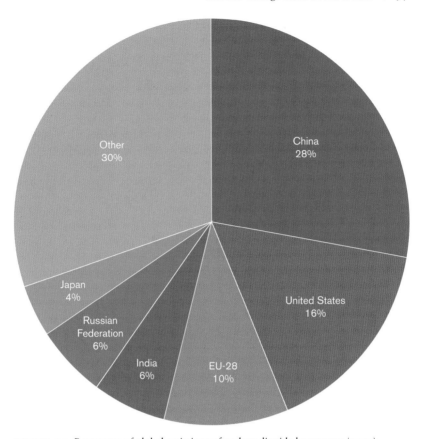

FIGURE 5.1. Percentage of global emissions of carbon dioxide by country (2015).

land, primarily in vegetation. While we are also emitting methane and nitrous oxide, $CO_2$ makes up 76 percent of greenhouse gases.

Global temperature is directly responsible for the Earth's climate, which in turn determines how much of the planet's water is locked up in ice sheets and glaciers and how much is contained in the oceans as salt water. Global temperature also affects the temperature of the ocean and how much volume it takes up. As the planet warms, as it has been doing since the last ice age ended about 18,000 years ago, ice has been melting and the oceans have been gradually warming, with both processes conspiring to slowly raise sea level. At the end of the last ice age about 10 million cubic miles of ocean water was tied up as ice sheets and glaciers, which lowered sea level nearly 400 feet. The subsequent postglacial warming raised sea level in fits and starts to its present elevation.

## SEA-LEVEL CHANGE: DRIVING FORCES

The climate and temperature of Earth are influenced primarily by the amount of heat it gets from the sun, which is directly related to Earth's distance from the sun at any period in time. The Serbian engineer Milutin Milankovitch recognized nearly 150 years ago that the distance between the Earth and the sun changes over time because of irregularities in the Earth's orbit, as well as the tilt and wobble of the Earth on its axis (figure 5.2). As the axis of the Earth's rotation tilts a few degrees, or its orbit moves it farther away from the sun, we receive less solar energy, and as a result the Earth and its atmosphere and oceans cool slightly. When the Earth is closer to the sun, global temperatures increase. There are well-understood cycles in these orbital oscillations that span tens of thousands of years and are associated with well-documented warming and cooling intervals that have profoundly affected the Earth's surface, its ice cover and the ocean volume and therefore the location of the shoreline. These orbital differences in distance are not huge but have been sufficiently large to lower or raise the Earth's temperature enough to help either initiate or terminate an ice age.

There are some important feedbacks that magnify the heating or cooling effects of these orbital variations. When the Earth begins to enter a warmer, or interglacial, phase, three important processes are affected. First, as more shelf ice or floating Arctic ice melts, more ocean surface is exposed to sunlight. The ocean surface absorbs heat, in contrast to ice, which reflects heat. Second, permafrost at high latitudes starts to thaw as the Earth begins to warm, and in doing so it releases carbon dioxide and methane from the decaying and formerly frozen vegetation it contains. And third, a warmer ocean cannot hold as much $CO_2$ as cold water, so as the ocean warms $CO_2$ is released to the atmosphere, adding to the greenhouse effect and causing more warming. Each of these processes can also move in a reverse direction, when the Earth starts to cool as a result of its relationship to the sun.

All other things being equal, and they rarely are, the warmer it gets, the more ice will melt and the higher sea level will rise. There is only so much ice on the surface of the Earth, however, so there is a limit to the extent sea level can rise from climate change. The mountain glaciers of the Earth, those in the Himalayas, Patagonia, the Alps, the Andes,

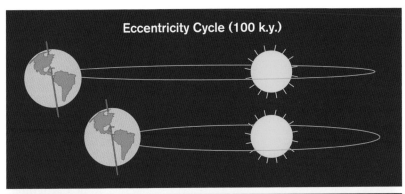

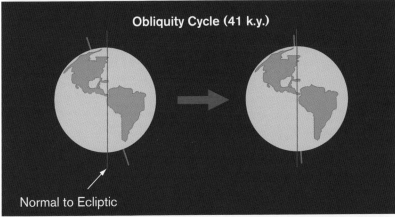

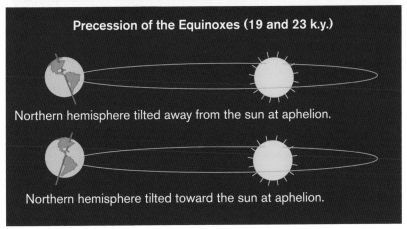

FIGURE 5.2. The three irregularities in the Earth's rotation and in its orbit around the sun, known as the Milankovitch Cycles (KY = thousands of years).

Alaska, and even Glacier National Park, while impressive in their own right, are relatively small in area and total volume. Virtually all of them have retreated significantly over the past 150 years. If these glaciers continue to retreat and were to completely melt, they would add only about 2 feet to the total level of the oceans. While relatively small compared to the two other large storehouses of ice, this could be very significant if you live within 2 feet of sea level, and lots of people around the planet do.

Greenland is an entirely different story, however. If its ice cover were to completely melt or calve off into the ocean, sea level would rise about 24 feet. Greenland is melting at an increasing rate (figure 5.3), and 24 feet of sea-level rise would inundate many densely populated areas around the world that are situated virtually at sea level. A number of large cities are already experiencing seawater in some of their streets now, in a process often called tidal or nuisance flooding. Along the Atlantic seaboard, these include Boston, New Haven, Atlantic City, Baltimore, Ocean City, Norfolk, Charleston, Savannah, and Miami, to name a few (figure 5.4). Future sea-level rise will increase the frequency of this wet inconvenience in very predictable ways (figure 5.5). Communities bordering San Francisco Bay also experience tidal flooding today, as do some Southern California beach communities. Globally, over 150 million people live in parts of cities within 3 vertical feet of high tide, including London, Alexandria, Kolkata, Mumbai, Dhaka, Yangon, Bangkok, Jakarta, Ho Chi Minh City, Hai Phong, Shanghai, Guangzhou, and Tokyo, to name a handful. This tidal flooding is discussed later in this chapter.

Antarctica has by far the largest volume of ice, about 70 percent of all the freshwater on the planet, and enough to raise sea level globally by about 200 feet were it all to melt. The ice shelves surrounding parts of Antarctica are melting at rates that are historically unprecedented, but the largest volume of ice in Antarctica lies in the interior of the continent, locked in by mountains, so it is not all likely to melt anytime soon. On the other hand, scientists who study Antarctica have serious concerns about the potential for one or more of the very large glaciers to accelerate and break off in the decades ahead, which could lead to a very rapid rise in sea level of several feet or more. The floating ice shelves are presently acting as buttresses or corks to keep the glaciers from advancing quickly. However, with continued ocean warming and sea-level rise, the concern is that these ice shelves will break up, allowing the very large glaciers to advance and calve off into the sea.

FIGURE 5.3. Meltwater river and moulin (a deep circular pit where meltwater enters a glacier) on the Greenland Ice Sheet, 2010. (Photo: Adam Scott © 2010, www.adamscottimages.com)

FIGURE 5.4. Nuisance flooding during high tides is becoming increasingly more common in many cities along the Atlantic Coast of the United States. At Annapolis, Maryland, in December 2012 wind, rain, and high tides combined to create significant flooding. (Photo: Amy McGovern licensed under CC BY 2.0 via Flickr)

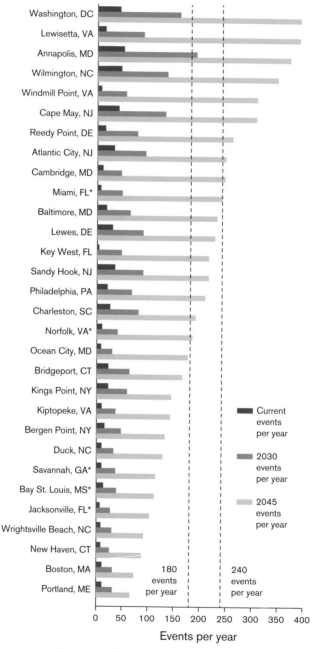

FIGURE 5.5. Frequency of tidal flooding in cities along the Atlantic Coast of the United States in 2012 and projections for 2030 and 2045. (Courtesy of Union of Concerned Scientists, *Encroaching Tides,* www.ucsusa.org)

While global cycles of heating and cooling, or interglacial and glacial periods, have been going on for several millions of years, these natural fluctuations have been altered in a major way during the past 150 years by the increasing concentrations of greenhouse gases in the atmosphere from human activities, primarily fossil fuel combustion (~91 percent of contribution) and land use changes such as the burning of tropical rain forests and other vegetation (~9 percent).

The Earth's atmosphere has always contained natural greenhouse gases, primarily carbon dioxide, methane, and nitrous oxide, and these have made the difference between a barely habitable planet and one where life as we know it can thrive. Without this natural greenhouse effect, the average temperature on the planet would have been about 0°F (−18°C) rather than the 60°F (15°C) we enjoy. The greenhouse effect is not a new idea; it goes back over a century to the work of the Swedish chemist and Nobel Prize winner Svante Arrhenius. But human activity has been adding to the natural greenhouse content of the atmosphere for well over a century, and these anthropogenic increases have been well documented.

The longest records of the changes in the greenhouse gas content of the atmosphere have been recovered in recent years from the ice of Antarctica. As snow falls and accumulates on the ice fields of the southern continent, air bubbles are trapped between the snowflakes and are then frozen into the ice. By drilling down through the ice, now as deep as 11,000 feet below the surface, continuous ice cores have been recovered that extend back over 800,000 years. The chemistry of these preserved air bubbles has been analyzed and reveals that atmospheric $CO_2$ concentrations fluctuated between about 180 and 300 parts per million (ppm) over this long period of prehuman history (figure 5.6) but have increased significantly over the past century.

Almost halfway around the world, on the flank of Moana Loa on the island of Hawai'i, daily measurements of the carbon dioxide concentration in the atmosphere document a 27 percent increase since the late 1950s. By 2016 the atmospheric $CO_2$ concentration in Hawai'i had reached 405 ppm (figure 5.7). This represents a 43 percent increase over the natural levels that persisted for hundreds of thousands of years that are preserved in Antarctic ice. There is no longer a scientific question that human activities have enhanced the natural greenhouse effect, leading to a warming of the Earth of nearly 1.5°F, and that this trend continues to increase. The warming of the Earth is causing the oceans to heat up and more ice to melt, which combine to raise sea level.

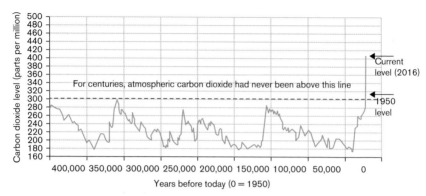

FIGURE 5.6. The carbon dioxide content of the atmosphere is now significantly higher than it has been at any time during the past 400,000 years as recorded in Antarctic ice cores. (Image: NASA)

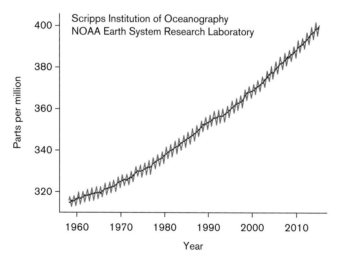

FIGURE 5.7. The carbon dioxide content of the atmosphere has been measured from the summit of Moana Loa in Hawai'i since the 1950s and has continued to increase. (Courtesy of Scripps Institution of Oceanography and NOAA)

## PAST AND PRESENT MEASUREMENTS OF
## SEA-LEVEL RISE

While we cannot directly measure where sea level was 100,000 or 1,000,000 years ago, there is a direct connection between global climate and sea level, so that proxies, or records of past climate, can be used to determine prior sea levels. These records include the fossils preserved in deep-sea sediments (which extend back over 60 million years), ice cores from Antarctica and Greenland (extending back over 800,000 years), tree rings (which extend back about 5,000 years in the case of California's bristlecone pines), as well as pollen records from lake sediments and growth rings in deep-sea corals. We have also collected seafloor samples from the continental shelves around the world, which have preserved the record of the past 18,000 years of sea-level rise. On the floor of the North Sea, for example, between the United Kingdom and France, peat samples have been dredged up and dated using carbon-14. Peat forms in freshwater bogs or swamps, often in coastal areas very close to sea level. If we know the depth below sea level where the peat was recovered and can determine its age through radiocarbon dating, we can place a point on the sea-level rise curve of approximately where the shoreline or sea level was at that point in time. Datable samples of fossil corals can also help us determine where sea level was at various times in the past. Many samples from continental shelves around the world have enabled us to develop a good record of sea level for this entire time period (see figure 1.1).

From the end of the last ice age approximately 18,000 years ago to about 7,000 years ago, sea level rose globally at an average rate of about 12 millimeters per year (mm/yr.), or 39 inches per century. Within this period of relatively rapid rise, however, there appear to have been pulses when rates were even higher for shorter intervals, 20 mm/yr., or 78 in./century, which are believed to have been due to major pulses of glacial melt from Antarctica. This is a very high rate relative to anything we have seen throughout the entire period of human history. Climate change slowed to a crawl about 7,000 years ago, and from geologic records of nearshore deposits recovered from the seafloor, it appears as though sea level rose at a very low rate, perhaps one millimeter per year, until the last 150 years or so.

The oldest tide gauges or coastal water level recorders were established in the mid-nineteenth century, and over the subsequent 150 years many others have been established along the world's coastlines. These gauges have continued to record the daily ranges but also long-term changes in sea level at each location, and a number of these now have histories of 75 to 100 years, which reveal a global average rate of sea-level rise for the past century of about 1.7 mm/yr. (~ 7 in./century). There is one major limitation to these tide gauge records, however. Because they are each anchored to some rigid mass of land, or some solid structure such as a bridge or wharf attached to land, they reflect the rise of elevation of the ocean relative to the land they are attached to.

In far northern latitudes where continents were depressed by thick covers of ice during the ice ages, there are many regions (e.g., Alaska, northern Canada, and Scandinavia) where the land is still rebounding from the ice removal (figure 5.8). Think about depressing a mattress when you sit on it and then how it rebounds or recovers when you stand up, only the rebounding of a landmass takes place very, very slowly. The land in these formerly glaciated northern latitudes is rising faster than the ocean such that sea level is actually dropping relative to the coastline. At Yakutat in southeastern Alaska, for example, sea level is falling at 17.6 mm/yr. (5.8 ft./century). Needless to say, the residents of Yakutat are not losing a lot of sleep over sea-level rise. The gauge in Furuogrund, Sweden, in the Gulf of Bothnia, shows a regular drop in sea level of 8.1 mm/yr. (2.7 ft./century) for the past hundred years. There are other areas scattered around the world, where the opposite is taking place. The coastline is subsiding, from either petroleum or groundwater withdrawals or from consolidation of organic-rich sediments, such that sea level is rising relative to the land at much higher than global rates. The Gulf Coast of the United States is a good example; at Eugene Isle, Louisiana, not far from New Orleans, sea level is rising at 9.6 mm/yr. (38 in./century; figure 5.8), over five times the global average. The Gulf Coast has serious challenges ahead with future sea-level rise. Venice, Italy, is perhaps the most iconic of all the sinking regions around the world: because of land subsidence, sea level is rising there at nearly double the global average (see figure 11.1).

These wide variations in relative sea-level measurements from fixed tide gauges scattered around the world's coastlines and the uncertainties about the accuracy of an average value due to the uneven geographic distribution of the gauges led to the implementation of a higher tech approach in the early 1990s. Several satellites were launched that use radar combined with GPS (Global Positioning Systems) to very precisely locate the positions of the satellites and their distance above the ocean surface in a process known as satellite altimetry. From 1993 to

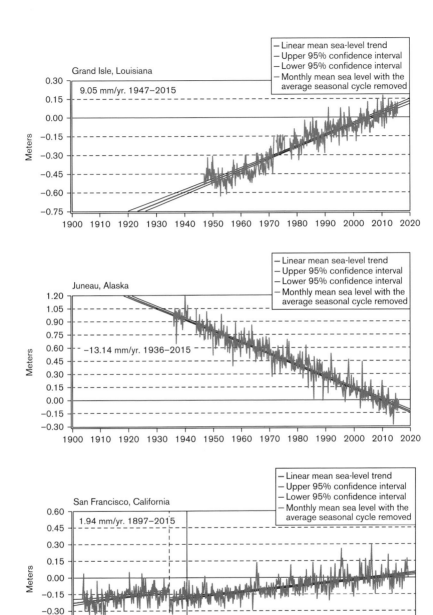

FIGURE 5.8. Sea-level rise values measured at tide gauges document local sea level relative to the landmass mass they are attached to. Alaska, San Francisco, and Louisiana tide gauge records illustrate these regional differences. (Courtesy of NOAA)

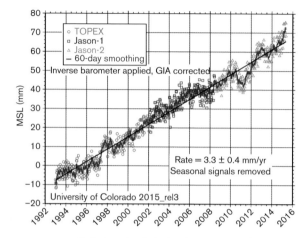

FIGURE 5.9. Global sea-level rise has been measured using satellite altimetry since 1993 and shows an average increase of about 3.3 mm per year. (Image: Steve Nerem, University of Colorado)

2016 the absolute rate of global sea-level rise determined from satellite altimetry has been calculated at ~3.3 (± 0.4) mm/yr., almost double the 1.7 mm/yr. for the previous century based on the averaging of global tide gauges (figure 5.9). This equates to 33 centimeters, or about 13 inches per century, and provides all coastal states and nations with a value that they can begin to plan around. It is also important for future coastal land use planning that regional tide gauges, if present, also be considered when there is significant tectonic activity and the land is either rising or sinking.

## GLOBAL WARMING AND FUTURE SEA-LEVEL PROJECTIONS

Over the past decade or so, the discussion of climate change and global warming has moved from the university seminar room and scientific conferences to city council chambers and the halls of Congress. For many, it initially seemed inconceivable that anything humans were doing could possibly have any significant effect on the atmosphere and temperature of the entire planet. Today, however, 97 percent of the climate science community believes that human activity has had a significant impact on climate change and the warming of the planet and its oceans, which is leading to increasing rates of sea-level rise. The questions coastal communities, states, and nations all need to face, if they haven't already, is what the future rate of sea-level rise will be, how it will affect their own region or jurisdiction, and how they are going to respond or adapt to the flooding, inundation, and shoreline retreat that will accompany the rise. In the United States, projecting population growth to 2100, there will be 4.2 million people living within 3 vertical feet of high tide. If sea level rises 6 vertical feet, the number of affected people rises to 13.1 million. With approximately 150 million people around the world today living within 3 feet of present sea level, for much of the population in large cities scattered around the world's coastlines, knowing how high sea level will be at various times in the future becomes an extremely important bit of information to know and use. Denial is not an effective strategy.

There are some coastal communities and areas that are already experiencing severe problems related to climate change and sea-level rise. One well-publicized group is the communities and island nations in the Pacific and Indian Oceans that occupy coral atolls, in most places just a few feet above sea level. The Maldives, Kiribati, Tarawa, and Tuvalu are just a few of the places where flooding and contamination of limited freshwater aquifers from a rising sea are endangering their very survival as communities and nations (figure 5.10). A foot of sea-level rise can make a difference when you live almost at sea level to begin with and there is no higher ground to move to.

At the opposite climate extreme are the Alaskan Inupiaq and Yu'pik villages along the Beaufort, Chukchi, and Bering Seas. Kivalina has become the poster child for these subsistence hunting villages, which typically are each home to perhaps three hundred to six hundred people. Kivalina is on a sand and gravel barrier island, historically underlain by permafrost, which remained relatively stable. Climate change–related processes are now threatening Kivalina and other similar villages. Although still fairly stable as far as barrier islands go, the formerly solid permafrost foundation has now begin to thaw seasonally, such that it is much more vulnerable to wave attack (figure 5.11). In addition, as the previous shore-fast ice, which seasonally armored the edge of the island, has melted and retreated each summer, there is more sea surface exposed to the wind, so larger waves can form. When the waves reach the shoreline there is no ice left to buffer their impact on the now weakened and thawing permafrost. In addition, thawing permafrost is more likely to lead to subsidence-related coastal flood hazards under increasing exposure to ice-free conditions. Each of these small villages is now facing a dilemma similar to that of the low-lying tropical Pacific islands.

FIGURE 5.10. This household on South Tarawa in the tropical Pacific is threatened by a retreating shoreline. Moving away from the shoreline is not an option as the island is very narrow and already crowded. (Photo: Government of Kiribati licensed under CC BY 3.0 via Wikimedia)

FIGURE 5.11. The low bluffs of Kivalina, Alaska, are increasingly susceptible to subsidence and erosion as the underlying permafrost thaws. (Photo: Nicole Kinsman, NOAA/NOS)

Projecting or predicting future sea levels is a difficult problem for many reasons. As a distinguished scientist once said, "Prediction is difficult, especially about the future." Yet there are scientists from many different nations, many of them directly involved with the United Nations Intergovernmental Panel on Climate Change (IPCC), who are using all of the tools in their toolboxes to make the best possible predictions. Projections of future global sea level are commonly made using mathematical models of the fundamental processes that contribute to global sea-level change, primarily the transfer of freshwater from melting ice (the *cryosphere*) to the oceans and changes in water density (typically labeled *steric* changes) arising mainly from the thermal expansion of ocean water as it warms. At present we believe that ocean warming is contributing about one-third of the rise, with ice melt making up the remaining two-thirds. The former is more straightforward because we know the thermal expansion of water at different temperatures, but it does involve some assumptions about how much individual layers or depths of the oceans' water will be warmed at various times in the future. The question of how much the melting of glaciers and ice sheets, primarily in Greenland and Antarctica, will contribute in the decades and centuries ahead, however, is much more complicated simply because as humans we have never before witnessed ice melt on this scale.

There is another approach to projecting future sea level, the semiempirical approach, which is based on the observed relationship between past sea levels and past global temperatures. This approach essentially combines all of the individual contributors and their importance into a single correlation between global temperature and ocean level. There are geologic records from deep-sea sediments and ice cores, from coastlines and the continental shelves of the world, which provide good indicators of past sea level and global temperatures. These have been compared to give us a reasonable connection between future temperatures and how much sea-level rise they may generate.

Virtually all future projections include ranges based on different greenhouse gas emission scenarios, from aggressive reduction of emissions to business as usual. The range of estimates using these methods and the available data diverge as we look further into the future, simply because the uncertainties regarding global greenhouse gas emissions get larger. The largest uncertainties are not with the science, however, but with the politics and social responses to climate change in the 194 nations around the planet, primarily the top six greenhouse gas emitters, China, the United States, the European Union countries, India, Russia, and Japan.

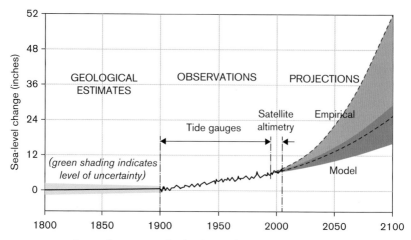

FIGURE 5.12. Past and present sea-level values and projections for the future out to 2100. ( NAS-NRC Sea-Level Rise for the Coasts of California, Oregon and Washington)

While new projections and revisions continue to appear, a 2012 U.S. National Academy of Sciences report is still being used as a reference guide for the western states (California, Oregon, and Washington; figure 5.12). The range of estimates for global sea-level rise from that report for 2050 lie between about 6 and 18 inches (15 and 45 cm), with a midpoint of about 12 inches (30 cm). For 2100, most ranges cluster between 20 and 55 inches (50 and 140 cm), with an average or midpoint of about 36 inches (90 cm). However, this will continue to be a moving target as long as greenhouse gas concentrations continue to increase and global warming persists, which is likely to be for at least several hundred years.

## ASSESSING FUTURE VULNERABILITY TO SEA-LEVEL RISE

Projections of future sea levels should be used as reasonable guides for what a community or region may expect in the future, but there is little doubt that these values will continue to be revised as new data become available. Any community, state, or nation planning for specific projects or proposals or developing long-term responses or adaptation strategies should also consider several factors:

(1) the sea-level rise value from the closest tide gauge, because there are regional differences due to land motion;

■ Area vulnerable to an approximate 16-inch sea level rise
■ Area vulnerable to an approximate 55-inch sea level rise

0 0.5 1    2 Miles ↑

FIGURE 5.13. Areas around central San Francisco Bay that would be inundated by a 16-inch and a 55-inch rise in sea level. (Courtesy of San Francisco Bay Conservation and Development Commission)

(2) the cost and lifetime of the facility being considered, whether a multimillion-dollar sewage treatment plant engineered to last fifty years or a city park or bike path;

(2) the impact of damage to or loss of the project or facility and the cost of replacing or rebuilding it.

The impact of sea-level rise or flooding on a large power plant or an international airport (figure 5.13), for example, will be of far greater consequence than damage or loss of a coastal parking lot or roadway, and for the former, a much more conservative value of future sea level should be used for planning and siting.

While we do need to think in the long term and about what the range of future sea-level rise projections are for any particular area of coastline, there are short-term water levels that are of much greater concern. Some of these have been discussed in chapters 2, 3 and 4, events like tsunamis, hurricanes, severe storms, and extreme high tides, but there are the more annoying and regular events that are becoming more frequent, especially in some of the low-lying Atlantic seaboard cities. Norfolk, Atlantic City, Charleston, and Boston are just four of the places where sea levels rose 5 to 9 inches (12 to 22 cm) between 1970 and 2012, while the global value was only about 3 inches (7.5 cm). In addition, as mentioned earlier, these cities now experience between six and almost thirty days each year of nuisance flooding, when seawater at high tide is inundating streets and neighborhoods with increasing regularity (see figure 5.4). Other Atlantic Coast cities, Wilmington, Washington, DC, and Annapolis, for example, are expected to experience approximately 130, 160, and 180 days each year, respectively, of nuisance flooding by 2030.

Along the California coast, "King tide" has become a common phrase in many low-lying communities. Several times each year, normally during the winter months, the irregular orbits of the Earth around the sun and the moon around the Earth conspire to produce these extreme high tides. In low-relief areas already very close to sea level, even a few inches can make a difference, especially when large storm waves arrive coincident with these extreme tides, giving us a glimpse of the future. The public has now gotten involved in documenting and publicizing these events on a social media website where images of flooding and very high tides are instantly uploaded, giving people a realistic image of what we are already experiencing in low-lying areas (figure 5.14).

Similar short-term causes of elevated sea levels are the periodic El Niño events, which affect the entire U.S. West Coast. During the large 1982–83 and 1997–98 El Niño events, the bulge of warm equatorial water that migrated up the West Coast elevated sea levels 12 to 24 inches (30 to 60 cm) above predicted tidal elevations. When combined with normal winter high tides and large waves, coastal flooding and erosion were severe. The 1982–83 El Niño inflicted over $240 million (in 2015 dollars) to oceanfront property in California when waves

FIGURE 5.14. The Embarcadero in San Francisco during a very high (King) tide. (Photo: Sergio Ruiz © 2012)

attacked beach-level or back-beach homes in places like Stinson Beach, Rio Del Mar, Aptos, Solimar, Faria, Malibu, Del Mar, Oceanside, and Imperial Beach. Damage was not restricted to broken windows and flooding of low-lying areas; thirty-three oceanfront homes were totally destroyed, and over three thousand homes and businesses were damaged, along with park improvements, roads, and other public infrastructure (figure 5.15).

## WHERE DO WE GO FROM HERE? RESPONDING AND ADAPTING

Sea-level rise has been recognized as a major threat to low-lying areas around the world since the issue of human-influenced global climate change first emerged in the 1980s. A growing number of reports and studies have illuminated and detailed the potential effects of a continuing and increased rate of sea-level rise. Many regions are already feeling the impacts, and there are many more that will in the decades ahead. Higher sea levels will erode beaches, dunes, bluffs, and cliffs and will

FIGURE 5.15. The 1983 El Niño damaged 3,000 oceanfront homes and businesses along the California coast and destroyed 33 homes. (Photo: Gary Griggs © 1983)

temporarily flood and then permanently inundate low-relief estuaries, marshes, lagoons, and wetlands as well as any associated development and infrastructure. The effects of a rising sea on coastal cities and nations across the planet may be the biggest threat that human civilization has ever faced. Some small island nations that are only a few feet above sea level are already feeling the impacts and have been making plans for relocation to safer sites. The bad news and the good news is that it is still happening somewhat slowly and has been overshadowed by the more extreme short-term events. But the rising sea is a ramp, with these extreme events, whether storm waves, El Niños, or hurricanes, all riding on top.

Coastal communities, cities, states, and nations must start planning now, if they haven't already begun, for how they are going to respond to the future rise in sea level, which is now guaranteed and unavoidable. The coastal environments present in any region (whether beach, wetlands, estuaries, dunes, bluffs, or coral atolls) will respond differently, but assessing the vulnerabilities of each area to future sea-level rise is an important starting point. What are the historical or documented rates of bluff, dune, or shoreline retreat? Are there accurate elevation maps so that we not only know which areas have been flooded in the past and how often, but also which ones are likely to be affected in the future? This is the type of information that is needed to begin to develop response or adaptation plans.

The options available are limited, and many of them should be recognized as relatively short-term fixes. Beach nourishment and armoring, whether seawalls or rock revetments, fall into that category. While they may work for the immediate future (i.e., the next several decades), depending on both the rate of sea-level rise and the frequency and magnitude of extreme events, we need to accept the reality that they are not going to be successful over the long term. Managed retreat, or the gradual and planned retreat from the shoreline, will almost certainly become the accepted and only feasible long-term solution by midcentury and beyond. This is not going to be popular anywhere, but all protection ends somewhere, and we need to begin to look beyond the next election cycle and how we are going to live with the inevitable rise in sea level.

# PART THREE

# Impacts of Human Activities on Coasts

CHAPTER 6

# Marine Pollution

*Domestic, Industrial, and Agricultural*
*Discharge*

As the number of human beings on the planet has risen, the population of coastal regions has climbed historically at an even faster rate. About half of the world's people now live within about 100 miles of a coastline. One of many benefits of living along the coast is that the ocean has always been a convenient place to get rid of waste, and communities and cities near the coast have always taken advantage of this (figure 6.1). Eight of the world's ten largest cities are located on the coast. When populations were much lower and waste products were less toxic and generated in lower concentrations, coastal waters were usually able to absorb and disperse the waste, through dilution, mixing, and biological breakdown.

As societies advanced and standards of living improved, advertising became a major industry and consumption rose. The chemical and pharmaceutical industries responded by developing literally thousands of new chemical products and compounds for many different uses, some beneficial, some not. Very well known examples of synthetic compounds are DDT (dichlorodiphenyltrichloroethane) and PCBs (polychlorinated biphenyls). The first proved to be very effective as an insecticide and in combating malaria but also had very detrimental effects on a wide range of birds. PCBs are fluids that have been used in plastics, as fire retardants, to strengthen wood and concrete, and to cool and insulate electrical transformers; they were soon recognized to have very detrimental effects on everything from plankton to marine mammals to humans.

FIGURE 6.1. A sewer outfall along the Central California coast formerly discharged primary treated wastewater 200 feet offshore in 6 feet of water prior to being connected to a secondary treatment plant with an outfall 2 miles offshore. The brown patch of water in the lower left surrounded by surfers is the effluent surfacing. (Photo: George Armstrong/California Boating and Waterways, collection of Gary Griggs)

These are just two of a long list of synthetic chemicals that were created for beneficial purposes but over time were recognized as being toxic to terrestrial and marine life as they were biologically magnified moving up the food chain. Yet cities on the coast, many of the larger ones being sites of heavy industry, have always used the adjacent ocean as a disposal site. The problems arise when the same populations who generate the waste also use the coastal waters for a wide variety of recreational activities and harvest and consume the fish and shellfish inhabiting them.

## SOME BASIC CONCEPTS

There are many types of contaminants or pollutants that may be constituents of domestic sewage, industrial effluent, or agricultural runoff, and their effects on nearshore waters are dependent on a large number of different factors. Among the most important are the nature of the contaminant or pollutant and its impact or toxicity to marine life, the concentration of the pollutant being discharged, and the circulation or mixing in the location where the effluent is released.

The initial concentration in the effluent is important, but once the pollutant is released, a number of other factors come into play that influence the effects or potential harm that the constituent can cause. These include the assimilative capacity of the receiving waters, or how much of the particular substance or chemical can be assimilated without causing harm or damage to the biota. This depends on the currents and circulation in the area of discharge, or how quickly the contaminant is mixed or diluted to a level that may no longer be harmful. It also depends on whether or not the particular pollutant is dissipated, broken down, biodegraded, or removed from the system. Heat from power plant cooling water, for example, may be dissipated fairly quickly in an area of active turbulence so that the area of concern may be relatively localized. Heavy metals like mercury or radioactive elements may persist for years, be concentrated by marine life, and remain harmful even in very small concentrations for decades.

Some pollutants may produce acute toxicity and be lethal to certain organisms relatively quickly. Heat can kill, and in places like Florida, where coastal water is already very warm, increasing the temperature 18°F in passing it through the cooling system of a large power plant can reach the thermal death point for some organisms. Other contaminants have more gradual or chronic effects, with the damage to the organism taking days, months, or longer to be felt, and then the impacts may be debilitating rather than lethal, causing metabolic disturbance, reproductive failure, or some other physiological impact.

Biological magnification, or the increase in concentration of a substance or chemical as it moves up the food chain, has been well documented and better understood in recent decades. DDT, discussed in a later section, is a good example. While low concentrations of some compounds at the base of the food chain may be harmless, as the chemical is magnified at each successive level, the impacts can become more and more detrimental and ultimately lethal. There are also synergistic impacts, or the combined or interactive effects of several different pollutants, heat and some particular chemical constituent for instance, where the impact of the substance is enhanced or becomes more detrimental at higher temperatures.

The biological impacts of any particular contaminant can be felt at a number of different levels and take different lengths of time to be recognized. At the most fundamental level, the impact may affect cellular or biochemical processes and be felt in minutes to hours. At a slightly higher level, effects may have an impact on the entire organism, and occur within hours to months. Some pollutants may affect entire populations or community dynamics and not be fully realized for months or years. At the highest level, there are global impacts, from perturbations like ocean acidification from carbon dioxide uptake by the oceans, which can take decades to be realized. We are now undertaking this very large and dangerous experiment.

## CONSTITUENTS OF WASTEWATER DISCHARGE AND RUNOFF

Unlike pollutants from large factories or industries, sewage from towns and cities belongs to us—we produced it—and like it or not, we own it. In the United States, many early sewage collection and disposal systems were just that, wastewater from homes, stores, schools, and small factories was collected in pipes and discharged into the most convenient body of water, whether a river, a lake, or the ocean. As the volumes increased and the problems and impacts of degraded water quality were recognized, treatment facilities were constructed, first in larger cities but eventually in nearly all communities of any size. With improved treatment, the quality of the effluent began to gradually improve, as did the quality of the receiving water. Finally, in 1972, the Federal Clean Water Act, which required secondary treatment for all communities nationwide, was passed and began to be enforced.

## ORGANIC WASTES, NUTRIENTS, AND DEAD ZONES

Organic waste is derived from domestic sewage, livestock feedlots, and a variety of industrial processes, including pulp and paper production, petroleum refining, and fruit, vegetable, and meat processing. While organic material is not actually harmful or toxic, breakdown by microbes in water requires oxygen, and depending on the amount and type of organic matter, it may demand a lot of oxygen, which means reduced oxygen content in the surrounding water. The demand of oxygen is usually called biochemical oxygen demand (BOD) and is a useful indicator of contamination by organic matter, whether from domestic sewage, industrial processes, or food production. Higher organisms, such as vertebrates, require greater levels of dissolved oxygen in the water for survival, whereas invertebrates require less. As oxygen levels decline from excess organic matter, fish will be the first to die off or disappear, followed by the invertebrates and finally the bacteria. The discharge of large volumes of wastewater with high BOD can be detrimental for animal life in coastal waters.

Nutrients, primarily nitrates and phosphates, are essentially fertilizers and encourage the growth of plants, whether corn, wheat, or rice on farms or microscopic algae in the coastal ocean. Depending on the land use within a drainage basin, the level of urbanization and agriculture, the source of many of the nutrients in the coastal ocean may be treated domestic sewage or river runoff. For the latter, where agricultural lands cover significant area, it is often in estuaries or areas near river mouths where the excess nutrients first accumulate and can become detrimental. In coastal waters, if these are diluted and mixed, they can actually be beneficial by fertilizing the base of the food chain, which can then be beneficial higher in the chain. However, where nutrients are discharged in high concentrations and in large volumes and where circulation is limited, excessive plant growth can occur. Eutrophication can result, which is the rapid proliferation of algae, which if not consumed by herbivores will require large amounts of oxygen to decompose. Excessive algae growth can often be observed as discolored water (red tides are one easily recognized example), but yellowish or greenish foams can also be easily observed along some shorelines in the late spring, summer, and even fall months. These blooms of plankton can overpower other marine life, leading in some cases to hypoxia, or the nearly complete loss of oxygen, which has been responsible for large fish kills and can also be detrimental to shellfish.

Decades ago farmers realized that the application of nitrogen fertilizer was a way to increase crop production and also allow farming in otherwise infertile regions. No organism, plant or human being, can survive without nitrogen, yet there is a shortage of available nitrogen globally. Seventy-eight percent of our atmosphere is nitrogen, but it requires some special plants to convert it to a usable form. These plants can "fix" nitrogen by breaking the nitrogen molecules apart so they are accessible to other plants. To expand the amount of nitrogen available for agriculture and increase crop production, however, modern scientists figured out how to make synthetic fertilizers, combining inert nitrogen from the air and hydrogen from natural gas. The philosophy that developed seems to have been that if a little nitrogen can improve farming productivity, then a lot must be even better. And many of the high-yield crops that were developed in the 1950s and 1960s required more nitrogen to reach their full potential.

Globally, on average, crops use less than one-half of the nitrogen that farmers add to the soils, with much of the excess applied in China, India, the eastern United States, Central America, and Europe. This is where the difficulties start, as nitrogen quickly goes from being a catalyst for greater crop yields to a contaminant that produces a whole variety of problems, from contamination of drinking water to excess plant growth in rivers, lakes, estuaries, and the coastal ocean that ultimately chokes fish and other marine life. Nitrous oxide, which is a greenhouse gas some three hundred times more potent than carbon dioxide, can ultimately end up in the atmosphere, contributing to global warming. Nitrogen is the gift that keeps on giving.

The term *dead zone* is applied to the more than 400 areas of hypoxia that have been identified around the world oceans, with over 160 recognized in U.S. waters alone. In these areas, most higher forms of marine life cannot survive due to low oxygen contents (figure 6.2). What is generally accepted as the largest dead zone lies off the mouth of the Mississippi River and extends from Louisiana to Texas. The average size of this area of oxygen-depleted water is about 5,400 square miles, equivalent to the state of Connecticut, although in 2002 it encompassed 8,600 square miles. Every spring and summer, nitrogen from agricultural fertilizers applied throughout the farms of the Mississippi River watershed wash down the river into the Gulf of Mexico, where they trigger a bloom of phytoplankton, which begins the process again. The Mississippi drains about 40 percent of the entire lower forty-eight states and carries over a billion tons of nitrogen in a typical year, much of it washed off agricultural fields in the midwestern farm belt. But nitrogen also comes from sewage treatment plants, animal manure, and industrial atmospheric emissions. Estimates are that the Gulf dead zone may be costing over $80 million annually in reduced tourism and low fishing yields.

Dead zones can also occur from natural processes. Off the coast of Oregon, 3,500 miles away, without the aid of fertilizers from river runoff, the state has been dealing with its own dead zone. This zone was first noticed off central Oregon in 2002 and traced back to an influx of cold nutrient-rich and oxygen-poor water that had migrated from the north. But in the following years, the natural process of upwelling brought additional nutrients to the surface, which led to dead zones each summer and produced large kills of bottom fish and crabs. This Pacific Northwest dead zone has been expanding north toward Washington State's Olympic Peninsula.

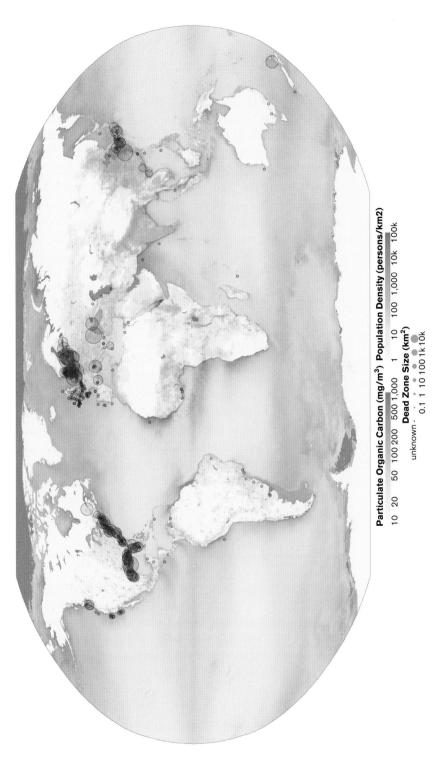

**Particulate Organic Carbon (mg/m³)   Population Density (persons/km2)**

10  20      50 100 200  500 1,000         1    10    100  1,000  10k  100k

**Dead Zone Size (km²)**

unknown ·    · · ● ●
0.1  1  10  100  1k 10k

FIGURE 6.2. Distribution of dead zones throughout the world's oceans by size in relation to population densities and amount of oceanic particulate carbon. (NASA)

On the opposite side of the country, the eastern portion of Long Island Sound, close to the densely populated New York City area, has experienced hypoxia and dead zones nearly every summer since at least the mid-1980s. While natural processes, such as upwelling, can produce dead zones, those in the Gulf of Mexico and in many areas are induced by excess nutrients from anthropogenic sources, whether agricultural runoff or domestic sewage.

Can these dead zones ever recover? The Black Sea in central Asia provides an example of what can be accomplished if concerted action is taken. This eastern arm of the Mediterranean had one huge hypoxic zone, extending over 15,000 square miles, or about 9 percent of the entire area of the sea. The use of fertilizers in the watersheds to the north and the associated runoff were reduced by over 50 percent when agricultural subsidies from the Soviet Union were terminated in the late 1980s. Although it took several years for the drainages to recover, combined with international assistance for runoff management, the Black Sea has recovered and remained healthy ever since.

Considerable research has been carried out on the pathways of nitrogen through the soils; the use of cover crops like fava beans, clover, soybeans, and alfalfa that can fix nitrogen without the use of chemical fertilizers; the use of moisture sensors and drip systems instead of furrow irrigation; and even the monitoring of nitrate in the soil. Many farmers have significantly decreased their use of both nitrogen fertilizers and water by adopting practices such as these and have dramatically reduced the runoff of nitrogen without reducing crop productivity. The methods and technology are there; they simply need to be introduced and implemented on a far larger scale. Education is crucial here.

## HARMFUL ALGAL BLOOMS

Although most of the microscopic plants or algae (phytoplankton and cyanobacteria) in the ocean are generally harmless, there are over twenty that produce toxins and can become harmful to animals higher up the food chain. The occasional proliferation of these types of phytoplankton, known as harmful algal blooms (HABs), have become more frequent in recent years and have also occurred in new locations (figure 6.3). This may be due to more nutrients being discharged from either domestic sewage or runoff from heavy use of agricultural fertilizers, or it may also be the result of more intensive monitoring efforts. While harmless to the shellfish and small fish such as sardines and anchovies,

FIGURE 6.3. Harmful algal bloom in Pinto Lake, Monterey Bay area, California. (Photo: Robert Ketley © 2005)

which feed on the plankton, these neurotoxins, as they are concentrated or accumulate higher up the food chain, can cause dizziness, memory loss, and even death in seabirds, marine mammals, and humans.

Twenty-five years ago HABs were relatively uncommon. This has changed in the intervening years, however, with these events now far more frequent and occurring in an increasing number of areas in coastal regions around the world. There is no scientific agreement on just why these events are so much more widespread, but terrestrial runoff and pollution, climate change, dispersal through ballast water, and more intensive observation and monitoring have all been implicated. As more and larger offshore regions are affected, more fish and shellfish industries have been impacted and economic losses have increased.

The consumption of shellfish, such as mussels and clams, is one of the common ways that harmful algal blooms and their toxins can affect human health. Shellfish monitoring programs exist in many coastal states, and signs are typically posted in areas where mussels and clams are routinely harvested at times when the toxins in the shellfish are usually at harmful levels. Beginning in the summer of 2015 and extending through the following winter, much of the West Coast shellfish industry, including the Dungeness crab and rock crab season, was shut down by the largest and longest harmful algal bloom ever documented, extending from California through Washington. This closure had a huge economic impact because Dungeness crab provides the greatest revenue of any fishery in Washington and Oregon and is usually the second most valuable in California. In 2014 the three states had a combined income of nearly $200 million from these valuable crustaceans. Well above average ocean temperatures over a very large offshore area, combined with high concentrations of nutrients in nearshore waters from seasonal upwelling, generated an unprecedented bloom of *Pseudo-nitschia,* a dinoflagellate that produces a potent neurotoxic, domoic acid. This single-celled algae was fed on by shellfish and smaller forage fish, such as sardines and anchovies, which were then consumed by marine birds and animals, with widespread paralytic poisoning, often with lethal effects for marine birds and sea lions. Economic losses for this Dungeness crab fishery were estimated in the millions of dollars, although by the fall of 2016 crabs were finally cleared for consumption along the entire West Coast.

Harmful algal blooms have become global problems with major public health and economic impacts. The precise causal factors for these blooms are not always clear, although higher water temperatures and the availability of nutrients, whether from land-based sources or upwelling, often seem to be the common denominators. A warming ocean is part of a much larger global climate change challenge, as is the natural process of upwelling. Where nutrients from land, whether agricultural, domestic, or industrial sources, are a catalyst, there is the opportunity to work toward reducing the discharge of these fertilizers.

PATHOGENS

Pathogenic, or disease-causing, microorganisms, primarily bacteria and viruses, are problematic constituents of domestic wastewater in many areas of the world. They are detrimental to water quality and hazardous to public health. Cholera, typhoid, hepatitis, and a number of other diseases are transmitted by water contact, which includes drinking, bathing, washing, or preparing food with contaminated water. If there is no sewage treatment system or if domestic wastewater has been inadequately treated, these microorganisms can enter groundwater, surface water, or coastal waters and then affect humans using that water. The magnitude of the risk from harmful bacteria or viruses in nearshore waters is related to the number of individuals who are already infected

FIGURE 6.4. Beach posted because of high bacteria counts in Monterey Bay, California. (Photo: Gary Griggs © 2016)

or carrying some microorganism, the existence of or level of sewage treatment in the particular city or community, and the location and use of the areas where the treated wastewater is released or discharged. There are occasional breakdowns or excess loads placed on wastewater treatment plants at certain times of the year, as well as ruptures of sewage transmission lines, during natural disasters for example, when raw sewage ends up flowing into the ocean, with negative consequences that include beach closures (figure 6.4).

Because sewage lines operate by gravity, many beachfront or back-beach developments have sewage collection lines running beneath the beach. These can be threatened or damaged during severe winter storms or hurricanes when there is significant beach erosion, which can undermine, expose, or rupture the pipes. Some beachfront homes still use septic tanks for on-site sewage disposal that may be located directly beneath the beach. With increased storm wave attack, beach scour, and a rising sea level, these sites present major challenges that will have to be resolved in the near future.

Because of U.S. Public Health Service requirements for municipal water quality in the United States, most major pathogens were removed from human contact through treatment plants many years ago, and community water systems are normally quite safe. However, with increasing numbers of international travelers, as well as immigrants from nations where certain diseases are still common occurrences, the opportunities are increasing for microorganisms to be transmitted to other regions where inadequately treated wastewater is discharged close to shore.

In communities where there is widespread recreational beach use, the ocean water is usually routinely tested for sewage or pathogen contamination. Coliform bacteria (*E. coli*), which are present in the intestinal tract of all warm-blooded animals, are not pathogenic or harmful themselves, but they are easy to detect, and their presence in the waters along the shoreline is commonly used as an indicator of domestic sewage contamination. This is a simple and inexpensive testing procedure and should be part of the public health program in any coastal community. When *E. coli* counts exceed water contact limits, beaches are normally posted as unsafe, and this still does happen in various locations in the United States.

## CHEMICAL CONSTITUENTS

A century ago the wastewater produced by a typical community, or a neighborhood in virtually any city in the world, was fairly benign. The constituents in that sewage would be broken down by microbes and absorbed or otherwise assimilated by the coastal ocean where they were discharged. Over time, as simple wastewater collection and discharge systems were improved with actual sewage treatment plants, the quality of the discharge to coastal waters improved as the constituents were removed.

The chemical industry has changed all of that, however, and there are now literally thousands of new chemicals that have been created in the laboratory for a wide variety of household, industrial, or agricultural uses that do not break down readily. While many of these compounds have benefited humanity, by conquering pests and disease and creating new and often useful products, they have not been without a price or problem. Many of these are now recognized as having detrimental impacts on marine life as well as humans. Forty years ago the Environmental Protection Agency had already classed 35,000 of those used in the United States as either potentially or definitely hazardous to human health. In addition to compounds created by the chemical industry

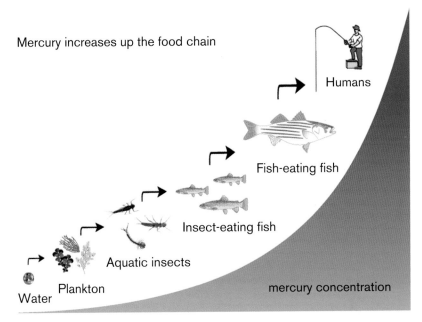

Mercury increases up the food chain

Humans

Fish-eating fish

Insect-eating fish

Aquatic insects

Plankton

Water

mercury concentration

FIGURE 6.5. Mercury undergoes bioaccumulation as it move up the food chain from plankton to higher organisms.

(e.g.,vinyl chloride, polychlorinated biphenyls, polycyclic aromatic hydrocarbons, trichloroethylene), there are also heavy metals and trace elements (e.g., arsenic, lead, mercury, cadmium) that occur naturally but are used in various industrial processes or products, and which all produce well-known public health problems.

In many cases, the effects of heavy metals or the impacts of new chemical products did not or may not show up immediately. It may take years after we have spread the chemical globally before the full impacts are recognized or can be traced to the specific pollutant. Mercury and lead are two of the most toxic metals used in the industrial world, and their impacts were recognized decades ago. Mercury is found in coal and is released to the atmosphere when coal is burned in power plants, where it can be spread globally and end up in the ocean (figure 6.5). This heavy metal has also been widely used in plastics, as fungicides for seeds, and in antifouling paints. The Minamata Bay, Japan, tragedy in the 1950s, when mercuric chloride from a plastic factory was discharged into a bay that was a source of fish and shellfish for the surrounding residents, was one of the most widely publicized incidents of metal poisoning. The inhabitants of the area suffered an epidemic of neurological disorders, kidney damage, and birth defects. Forty-one people died from ingesting fish and shellfish contaminated with organo-mercury compounds from the factory discharge, and many serious birth defects were produced. Mercury continues to be a constituent of concern in large pelagic fish (e.g., swordfish, tuna, shark) from years of accumulation. Mercury has also recently been detected in fog along the California coast, raising red flags about existing mercury and its transport pathways and accumulation. Mercury enters the United States via the upper atmosphere and then through rainfall, and its source is believed to be coal-fired power plants in Asia, particularly China. This heavy metal is a global problem, and like other pollutants in the atmosphere and ocean, it does not stop at national borders. The Minamata Convention on Mercury, which was adopted in 2013 by 128 countries, has the objective of protecting human health and the environment, but the burning of coal continues and the mercury content in rainfall in parts of central North America is still increasing.

Lead is right up there with mercury on the toxic metal list; it is known to cause brain damage, behavioral disorders, and hearing loss and can be lethal at high concentrations. It is particularly harmful to young children and can stunt brain development. While lead was used for years as a gasoline additive (tetraethyl lead), this compound was phased out beginning in the 1970s, although not before lead was distributed globally through the atmosphere. By 2012, 185 countries had stopped using leaded gasoline, and six others had planned to phase it out over the next two years. Thus, although leaded gasoline use has been dramatically reduced, lead is still widespread in the environment. It was found recently at high levels in fish in the Arabian Sea and at elevated levels in the Indian Ocean near Singapore.

Of the literally thousands of synthetic chemicals produced by modern industry, DDT and PCB are two examples of substances that were created for their widespread use and benefit but had unintended negative impacts on organisms and marine life on a global scale. Dichlorodiphenyl-trichloroethane, or DDT, was the miracle chemical that was going to eradicate the mosquito that spread malaria, a disease that some scientists believe might have led to the deaths of as many as half of all humans who ever lived. First introduced in the 1940s and 1950s, DDT was widely used during World War II for malaria control in the South Pacific and also to control the spread of typhus in Europe. The chemical was helpful in the final eradication of malaria in both the United States and Europe, although by that time the mosquito and its breeding grounds had nearly been eliminated by other methods. Because of its properties as an insecticide, farmers adopted it quickly as a general-purpose and agricultural spray. However, DDT possessed a number of properties that proved to be biologically detrimental: it was resistant to breakdown, it evaporated with water and therefore was transported globally by wind, and it accumulated in the fat of organisms.

Through a lot of careful scientific research and detective work it was soon discovered that the pesticide was being concentrated up the food chain through a process known as *biological magnification*. The pesticide made its way from fields to streams and finally to coastal waters, where plankton, fish, and then marine birds accumulated progressively higher concentrations in their fat (figure 6.6). Ultimately it was the reproductive failure of brown pelicans and peregrine falcons that led to the banning and elimination of the insecticide's use in the United States and gradually in 170 other countries as well.

PCBs had a similar history, with a broad array of industrial uses but widespread biological impacts in the coastal ocean, ranging from reduced growth rates in plankton to behavioral changes, weakened immune systems, hormone disruption, reproductive failure, and death in marine mammals. Although the toxic effects were first noticed in very low concentrations in the 1930s, no manufacturing ban occurred until 1979, almost fifty years later.

## WASTEWATER TREATMENT: REDUCING POLLUTANT DISCHARGE

The challenge today for every coastal community, region, state, or nation is to control the discharge of all potentially hazardous materials through the treatment of wastewater to a level adequate to remove or reduce the concentration of potentially harmful constituents to acceptable levels, or otherwise render the substances harmless through the treatment process. With increasing coastal concentrations of people producing larger volumes of wastewater, which often contain a growing array of new chemical constituents, the challenges are significant.

As discussed earlier, the list of potentially problematic constituents in domestic sewage, and in industrial or agricultural discharges or runoff, is large and includes organic matter, nutrients, pathogenic microorganisms, and a wide variety of chemical constituents. As legislated by the

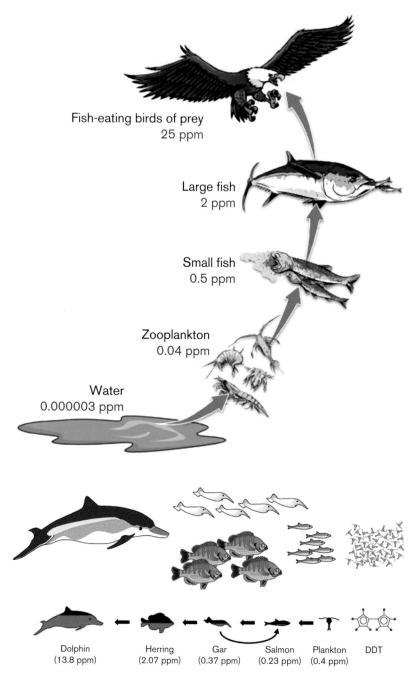

Fish-eating birds of prey
25 ppm

Large fish
2 ppm

Small fish
0.5 ppm

Zooplankton
0.04 ppm

Water
0.000003 ppm

| Dolphin | Herring | Gar | Salmon | Plankton | DDT |
|---------|---------|-----|--------|----------|-----|
| (13.8 ppm) | (2.07 ppm) | (0.37 ppm) | (0.23 ppm) | (0.4 ppm) | |

FIGURE 6.6. The pesticide DDT is concentrated in fat and undergoes biological magnification as it moves up the food chain to be concentrated in marine birds and larger fish.

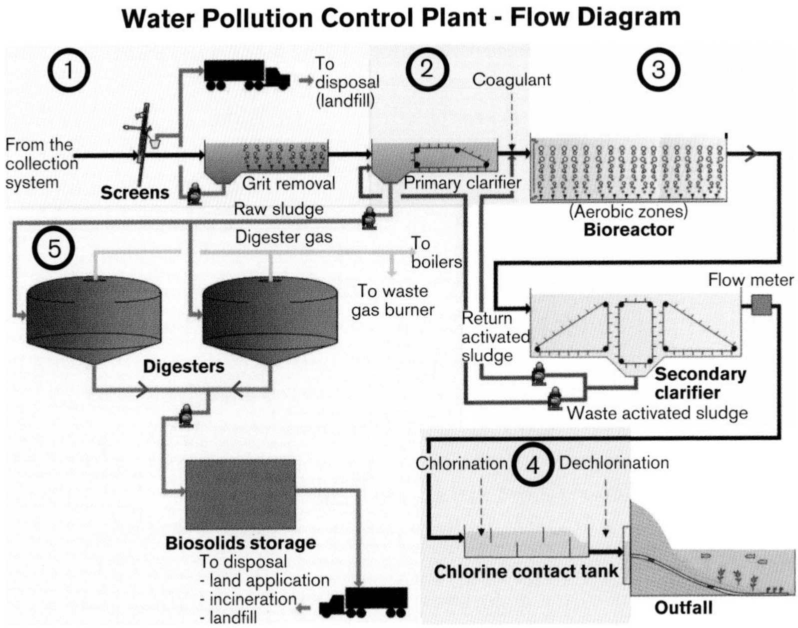

## Water Pollution Control Plant - Flow Diagram

FIGURE 6.7. Domestic wastewater in most coastal communities undergoes primary and then secondary treatment.

U.S. Clean Water Act of 1972, all domestic sewage discharges are supposed to undergo secondary treatment, although this is not the case in many other coastal nations around the globe.

Primary treatment, which is the first and most fundamental level of sewage treatment, essentially involves settling in large tanks, usually for at least several hours (figure 6.7). The material that settles or floats to the surface is removed and sent to a sludge digester, where it is heated and rendered more or less inert. This material may be disposed of on land or, sometimes, offshore. The remaining fluid is disinfected with chlorine, ozone, or some other agent and then discharged to a water body such as a river, a lake or estuary, or the ocean. While relatively simple, this process removes large fractions of the suspended solids and the accompanying biochemical oxygen demand, nutrients, metals and toxins, and pathogenic organisms. Disinfection of the treated water can be reasonably effective, although how far the effluent is discharged off-shore, the circulation and oxygenation in the receiving water, and the volume and constituents of the discharged fluid are all variables that determine the water quality of the surrounding ocean area.

Secondary treatment takes the primary effluent, after it has been through the settling process, and, using one of several different methods, puts it through a bio-oxidation process (see figure 6.7). These methods include a trickling filter, where the effluent trickles slowly through a tank open to the air that is filled with rocks covered with microbial life; an activated sludge tank, where bio-oxidation, or biological breakdown, occurs in a closed tank; or large, outdoor oxidation ponds, where algae and microbes break down and remove the remaining contaminants and constituents. After undergoing bio-oxidation, the settling process is again used to separate any remaining material that will float or sink, and that material is sent to the sludge digester for heating and breakdown. The fluid effluent is disinfected and discharged to a receiving water body, an estuary, a bay, or the ocean. Secondary treatment is effective in removing the great majority of the constituents discussed earlier, although many pharmaceuticals and other potentially harmful chemicals are not completely removed and have been detected in coastal waters even after this higher level of treatment.

Ocean waters along U.S. coastlines are far cleaner than they were fifty years ago because many industries have been forced to clean up their discharges, and sewage treatment plants have gone to secondary treatment and even more advanced treatment in some cases. There are still many chemical constituents that are not removed by even secondary treatment that have been detected in fish and other marine organisms and that can cause physiological and other problems. Regular monitoring of outfall and adjacent ocean areas is carried out in many locations, especially where ocean recreation brings large numbers of people to the shoreline and economic benefits are large.

Southern California is home to over 19 million people, and they use the beaches a lot, but they also discharge treated wastewater into the offshore area. The City of Los Angeles discharged raw sewage into Santa Monica Bay until 1925, when beaches became contaminated as a result of population growth and the discharge of more sewage. The first simple treatment facility was built, but it involved only screening of the wastewater and therefore did little to improve water quality. Twenty-five years later the city completed the Hyperion Treatment Plant, which used secondary treatment and also produced fertilizer as a by-product. As the volume of sewage continued to increase, treatment levels were modified in 1957 to handle the larger volume, with a mix of primary and secondary treatment and ocean discharge through a pipe that extended 5 miles offshore. With time, fertilizer production was terminated, and the digested sludge was discharged into Santa Monica Bay 6 1/2 miles offshore. Thirty years of sludge discharge led to almost the complete loss of benthic, or bottom, life in the bay, however, and monitoring revealed a failure to meet discharge quality standards. Through major upgrades to the facility at a cost of $1.6 billion, sludge was no longer discharged, and the 350-million-gallons-per-day plant now meets all water quality standards, and the beaches are safe.

## LOOKING TO THE FUTURE: WHERE DO WE GO FROM HERE?

Increasing numbers of people are moving to coastal cities globally, putting additional stress on coastal waters where the effluents from factories, farms, and cities ultimately end up. The evidence is clear that the effects from many of these discharges are increasing, whether larger and more frequent harmful algal blooms, expanding anoxic, or dead zones, heavy metals in pelagic fish, disease transmission, or the incidence of new pharmaceutical products in nearshore waters and marine organisms. The wealthier industrialized nations have greatly improved their discharge quality in recent decades, leading to safer coastal recreational conditions and high-quality seafood. However, poorer, less developed countries, particularly those in Africa, Latin America, and South Asia, have not yet developed the infrastructure to collect and treat the increasing volumes of wastewater being generated.

These are big issues that must be confronted directly by government agencies, as only they have the authority to implement and enforce improved water quality standards and discharge requirements. This requires political will from those in power, and the connections need to be recognized between the quality and constituents of ocean discharge and issues of public health, safety of the shellfish and fishing industries, and income from tourism and recreation. Greatly improved wastewater collection and treatment systems, longer offshore outfalls, and water quality monitoring programs are necessary starting points. We cannot use the same nearshore waters for our sewer, our recreational resource, and our fishing grounds without seeing harmful impacts.

There are some encouraging examples where the application and use of individual chemicals such as DDT, PCBs, and lead was finally terminated after the harm to humans and marine organisms was recognized. The U.S. Public Health Service requirement to treat domestic sewage to the secondary level before ocean discharge was also a significant step forward for improving coastal water quality. There is really only one big global ocean, and many of the constituents added from one nation will ultimately reach the shoreline of a distant one. The issue of improved coastal water quality has to be elevated to the appropriate government entity in all countries with coastlines and become a priority for the sake of the human and ecosystem health of each of these nations.

# CHAPTER 7

# Plastic and Other Marine Debris

## INTRODUCTION TO PLASTIC PRODUCTION AND CONSUMPTION

Our lives revolve around plastic today. But the word *plastic,* which is derived from the ancient Greek *plastikos,* meaning "capable of being molded or shaped," did not come into common use until 1907, when the world's first synthetic plastic was invented in New York City. The unique and useful properties of plastics—lightweight, flexible, strong, durable, moisture resistant, and relatively inexpensive—led to the rapid proliferation in the development and use of new plastic products throughout the twentieth century.

While some of us may have nearly eliminated plastic shopping bags and plastic water bottles from our lives, there is a long list of other common plastic products that we use every day, many more than most of us probably realize (box 7.1). Other than a few plastic products that we hope will last forever (hearts and heart valves and joint replacements, for example), most of this stuff is used just once (things like soft drink and water bottles, plastic bags and food containers, plastic cups and tableware, milk jugs and detergent containers) and then is recycled or tossed.

Industry produces more plastic each year, estimated at about 360 million tons globally in 2015. The amount is increasing at about 4 percent annually, and we consume it all. This is a difficult number to comprehend, and we probably need to think in other terms if it is to have any impact on our consumption habits. Relating plastics to the ocean

## Box 7.1 Types of Common Plastics

- *Polyester:* fibers and textiles
- *Polyethylene terephthalate:* soft drink bottles, films, microwave packaging
- *Polyethylene* (PE): bottles and bags and many other inexpensive uses
- *High-density polyethylene:* detergent bottles and milk jugs
- *Polyvinyl chloride* (PVC): plumbing pipe and rain gutters, shower curtains, window frames, flooring
- *Polyvinylidene chloride* (PVDC): food packaging (Saran Wrap)
- *Low-density polyethylene* (LPDE): outdoor furniture, siding, floor tiles, clamshell packaging
- *Polypropylene* (PP): bottle caps, drinking straws, yogurt containers, appliances, car bumpers
- *Polystyrene* (PS): packing foam (peanuts), food containers, tableware, disposable cups, plates, CD containers
- *High-impact polystyrene* (HIPS): refrigerator linings, food packaging, vending cups
- *Polyamides* (PA) (nylons): fibers, toothbrush bristles, tubing, fishing line, car engine parts
- *Acrylonite butadiene styrene* (ABS): computer monitors, printers, and keyboards; drainage pipes
- *Polycarbonate* (PC): compact discs, eyeglasses and lenses, security windows, traffic lights, riot shields
- *Polycarbonate/acrylonitrile butadiene styrene* (PC/ABS): car interiors and exterior parts; mobile phone cases
- *Polyurethane* (PU): cushioning foams, thermal insulation foams, printing rollers

may be the most meaningful way to contemplate the magnitude of the situation we have created.

## MARINE DEBRIS AND PLASTIC IN THE OCEAN

All of those properties that make plastic such a wonderful material for so many uses are precisely the same properties that make it a major concern for the oceans of the world. A casual walk on the beach, virtually anywhere on the planet, will bear this out. Coastlines from Alaska to

FIGURE 7.1. Plastic on an isolated beach on the Andaman Islands, Indian Ocean. (Photo © 2016 SAF—Coastal Care, Coastalcare.org)

Antarctica, from Mexico to Morocco, and from Taiwan to Tahiti will have their own collection of plastic debris, which may have traveled a few feet or a few thousand miles from its original disposal site (figure 7.1).

Let's look at just two of the most widely used and frequently discarded plastic products, bags and water bottles. It is now believed that shoppers around the planet use about 500 billion single-use plastic bags every year. This translates into a million bags being used every minute and nearly 70 bags a year for every person on Earth. If you lined up these bags end to end, they would stretch around the world at the equator nearly 5,000 times. No matter how many tons they weigh, that's an enormous number of plastic bags ending up somewhere.

How about plastic water bottles? While over a billion people around the world do not have access to safe drinking water, they usually are not the ones consuming all of the bottled water. Most countries where bottled water usage is highest have very good quality tap water. The irony is that 40 percent of all the bottled water consumed in the United States is simply filtered tap water, and it does not have to adhere to U.S. Public Health Service standards for clean water. In spite of this and for a variety of reasons (convenience, unfounded suspicions about tap water quality, and cute little bottles from Fiji or some other exotic place), Americans, who are almost all fortunate enough to have very good quality and regularly tested tap water, consumed 9.7 billion gallons of bottled water in 2012. This water was guzzled from 103 billion plastic bottles. This is 3,250 bottles emptied every second, all year long. If these bottles sold for just one dollar apiece, Americans spent almost $200,000 a minute on bottled water. Despite the prevalence of recycling programs and convenient containers for disposal, only 20 percent of the bottles are recycled; the rest are discarded and end up in landfills, on the beach, or in some cases in the ocean.

Looking at the global picture, people around the planet consume about 53 billion gallons of bottled water every year, supporting a $100 billion business. In convenient 12-ounce bottles, this amounts to about 560 billion plastic bottles every year and growing. If lined up end to end, they would extend around the world at the equator 3,500 times. In the United States alone, consumption of bottled water has been increasing at over 6 percent per year.

The amount of plastic that ends up in the ocean each year and where it comes from are difficult to determine, simply because no one is keeping track. Some scientists are trying to get a handle on how much gets recycled, how much ends up in landfills, and, for those populations living near coasts, how much of the rest probably ends up being discarded and may well end up in the sea. Looking at the concentrations of plastic containers and bags either on beaches or in coastal waters around the world certainly gives the impression that the numbers are huge. One recent study calculated that between 5 million and 14 million tons of plastic end up in the ocean every year. If we are ever going to have an impact on plastic reduction, we need to know where it is coming from and work to stop it at the source. The short answer is that it is coming from almost everywhere, but much of it comes from Asia. The largest single source is China, with an estimated 2.4 million tons per year, followed by Indonesia (975,000 tons), the Philippines (525,000 tons), Vietnam (500,000 tons), and Sri Lanka (425,000 tons). The United States ranks twentieth on the global list.

Unlike most treated wastewater, which is discharged through an outfall pipe, plastic gets into the ocean through a whole variety of nonpoint sources, making it much more difficult to control: direct discarding at the beach or along the shoreline; overflowing trash cans; transport downstream in a river, stream, or estuary; wind; and intentional disposal at sea from ships, to name some of the obvious mechanisms. In 2015 alone, an estimated 412 tons of plastic accumulated on the shoreline of Japan every day, and 300 tons ended up daily on the beaches of India (figure 7.2).

FIGURE 7.2. Plastic bags and containers cover the beach in Kochin, India. (Photo: D. Shrestha Ross © 2014)

Midway Island got its name for a reason; it lies almost in the middle of the North Pacific, about 2,400 miles from Japan and 3,000 miles from the west coast of the United States. But this mid-ocean location unfortunately has not spared it from being the recipient of plastic from thousands of miles away. The beaches on Midway are covered with plastic debris of every type, size, and color imaginable—and also with dead birds. Many of these birds were once graceful Laysan albatrosses, who ride ocean winds for thousands of miles but come here to breed and nest. Unfortunately, they are attracted to and consume the brightly colored pieces of plastic, thinking that this is food. The plastic is deadly, however, clogging their digestive systems, and many of the once magnificent birds are now just corpses of feathers and plastic where their stomachs had been (figure 7.3). It is estimated that one of every three albatross chicks dies from plastic ingestion. The Midway birds make up a significant portion of the estimated one million seabirds that die each year from plastic consumption.

Midway is not unique. Easter Island, in the South Pacific, is 1,800 miles from the coast of South America, yet its beaches are also littered with plastic. The same goes for Goa, a once romantic getaway island off the coast of India. At the opposite end of the Earth, Iceland is generally considered to have one of the cleanest shorelines on the planet, yet plastic

FIGURE 7.3. Albatross frequently die from plastic ingestion. Shown here Midway Atoll. (Photo: Ryan DeGaudio © 2006)

has found its way there also. The accumulation of plastic and other debris on island beaches is not a new phenomenon. When I visited the Greek Islands in 1975, over forty years ago, plastic detergent bottles and Styrofoam were scattered along the high tide line of every island I visited.

The term *marine debris* has been used to describe all of the waste that humans have discarded that ends up in the ocean. It includes a massive variety of plastic products (bags, beverage bottles, tableware and food containers, six-pack holders, detergent containers, lighters, and more) but also fishing floats and nets, clothing, shoes, lightbulbs, and anything else that will float. On the shoreline, cigarette butts are typically a major component as well. It is believed that 60 to 80 percent of all marine debris in the ocean is plastic, and it is found on the shoreline of every continent, and from Iceland to Antarctica.

## THE GREAT PACIFIC GARBAGE PATCH

In 1997 Charles Moore was sailing home with his crew from Honolulu to California after completing a trans-Pacific sailboat race. They took a northerly route, a shortcut, across the less traveled North Pacific Subtropical Gyre, where winds are generally calm and not much life exists. As they crossed this large stretch of ocean, they realized that they were surrounded by plastic. In the week that it took the sailboat to cross the gyre, they rarely saw a patch of ocean without plastic. Shampoo and detergent bottles, plastic bags and water bottles, Styrofoam and fishing floats, as far as they could see in the middle of the North Pacific. One of the crew members on the voyage called this "the Great Pacific Garbage Patch." The name stuck and quickly brought global attention to the issue of plastic and other debris in the sea and the impacts this was having on marine life.

Currents in all of the world's oceans flow in gyres, or circular patterns, clockwise in the northern hemisphere and counterclockwise in the southern hemisphere, and floating material tends to concentrate within them (figure 7.4). Ten years before Moore's voyage, in the mid- to late 1980s, a group of scientists from Alaska found large amounts of plastic and other debris that they believed were being concentrated by ocean currents. Fishermen and sailors had noted the debris in the past as well. In 1998 a publication by the National Oceanic and Atmospheric Administration (NOAA) described the high probability of the existence of "a large area concentrating plastic waste debris in the North Pacific." But it was not until Moore and his crew sailed through the Great Pacific Garbage Patch and held a press conference when they returned to their homeport in California that this was picked up by the media. The patch was initially described by the news outlets as "larger than the state of Texas," which gave the impression that the plastic literally covered the entire sea surface in these areas. It was even briefly labeled the "eighth continent." Several subsequent oceanographic voyages sailed through the area and sampled the plastic and other debris. Oceanographers discovered that there wasn't a big island or mound of plastic floating on the surface, but they recovered debris in over a hundred consecutive net tows across 1,600 miles of sea surface. Most of the plastic was in tiny pieces, a few millimeters across, like confetti, and most of it actually was drifting beneath the surface.

Additional research has shown that not only has much of the plastic, Styrofoam, and other debris broken down into small pieces, but as it degrades it is also releasing an entire cocktail of constituent chemicals that can be hazardous to marine organisms. Some of the debris is adsorbing or concentrating other contaminants, such as PCBs and DDT, in seawater. This makes these particles more dangerous when consumed by animals throughout the food chain: fish, turtles, marine birds, seals, sea lions, and whales.

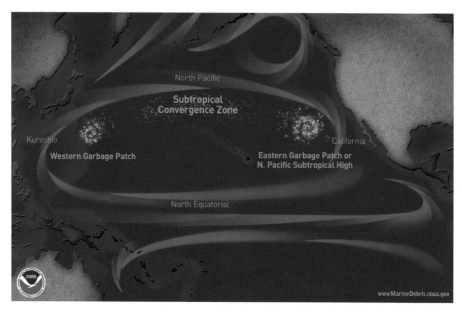

FIGURE 7.4. General locations of the two large garbage patches in the North Pacific. (NOAA)

With additional sampling and observations, it has now become evident that the North Pacific Subtropical Gyre actually consists of two separate debris fields, the Western and Eastern Pacific Garbage Patches (see figure 7.4). These stretch for hundreds of miles across the North Pacific, one lying between Japan and Hawai'i and the other between Hawai'i and California. However, these are not fixed masses of debris, nor are their boundaries clear and well defined. As wind and currents shift, the debris moves and circulates, accumulating more toxins and gradually breaking down, with some of it slowly sinking to the seafloor.

With similar circulation patterns in the other oceans, the South Pacific, the Indian, and the North and South Atlantic, it would seem to be just a matter of time until concentrations of plastic and other marine debris would be discovered in these areas as well. Surprisingly, a North Atlantic garbage patch was documented nearly forty-five years ago, in 1972, and was estimated to be hundreds of miles across, although shifting over time. For over two decades the Sea Education Association has been collecting and documenting debris from the patch with fine-mesh nets. But there was no name given to it, nor, apparently, was any press conference held, so it remained more or less below the radar.

FIGURE 7.5. Humpback whale in Hawai'i entangled in twenty-two different line and net types. (Photo: Ed Lyman, NOAA Hawaiian Islands Humpback Whale National Marine Sanctuary, Permit #932–1489)

In 2011 a team of scientists from the 5 Gyres Institute led a sampling voyage across the South Pacific from Robinson Crusoe Island off the coast of Chile to Pitcairn Island nearly 3,000 miles northwest, sampling the water about every 50 miles. Plastic concentrations were high in their samples, and they reported yet another oceanic garbage patch.

## THE CASUALTIES OF PLASTIC AND OTHER MARINE DEBRIS

From microscope plankton to very large marine mammals, it appears that all are being affected in some way by the plastic and other debris that continues to enter the ocean. A report published in 2006 lists 267 different species of marine animals that had suffered from either ingestion of plastic debris or entanglement in fishing nets or lines, six-pack rings, or other discarded debris. Seals, sea lions, dolphins, whales, sharks, turtles, and a wide variety of seabirds have all been caught or entangled, falling victim to cuts, infections, gradual starvation, or even drowning (figures 7.5 and 7.6). Accurate worldwide numbers are difficult to come by, simply because of the areal extent of the world oceans,

FIGURE 7.6. Blackfooted albatross entangled in string and balloons. (Photo: Jim Harvey, Moss Landing Marine Laboratories)

129 million square miles to be exact. But one estimate is that 100,000 marine animals die from plastic entanglement every year.

One of the most ubiquitous throwaways, plastic bags, seems to be found everywhere, and for marine animals like turtles, which feed dominantly on jellyfish, the bags look a lot like jellies from their underwater perspective. Choking or starvation can result when the bags are ingested. A Minke whale washed up onto the Normandy coast of France in 2002, and a necropsy revealed 1,760 pounds of plastic bags in its stomach. Fifty to 80 percent of dead sea turtles have been found to have ingested plastic debris. Bright-colored plastic like bottle caps, lighters, and a host of other durable items, whether in the water or along the shoreline, are mistaken for food by many seabirds and fish. Certain colors, particularly red, are more often consumed, perhaps because of the similarity in appearance to real food, like shrimp or fish eggs. Once swallowed, these plastic pieces can get stuck in windpipes, causing suffocation, block the digestive tract, or fill the stomach, leading to starvation or malnutrition.

What is not clear yet is whether the diverse set of chemical compounds that are contained in the ingested plastics or other debris, whether from breakdown of the plastics themselves or from uptake of DDT, PCBs, or metals from seawater, are taken up by the marine animals that consume them, creating additional health issues. This is an issue that requires more investigation.

## ECONOMIC IMPACTS OF MARINE DEBRIS

Beaches covered with plastic and other debris are not places where visitors or tourists want to relax and spend money (see figures 7.1 and 7.2). With tourist income such an important part of the economies of many island or tropical nations, this loss can be very significant. Tourism is the largest source of private capital coming into the Hawaiian Islands, and the 8.6 million visitors contributed $15.3 billion to the economy in 2015. Economic costs of marine debris on beaches include loss of tourism, as well as the expense of litter removal and waste management. One community on the west coast of Sweden, a nation that prides itself on being clean, spent over $1.5 million in a single year on beach cleanup. A small coastal community in Peru has calculated that it would have to spend about $400,000 annually to clean its coastline. Some California coastal communities spend tens of thousands to hundreds of thousands of dollars every year to clean up beaches and waterways. Offshore, the economic losses also include propeller fowling on ships and damaged engines, as well as reduced and lost fish catch.

A number of organizations and individuals have proposed cleaning up the Pacific Garbage Patch, an interesting proposition that seems like a good idea, although a whole lot more challenging than doing a weekend beach cleanup. There are a lot of misconceptions about that garbage patch, including its size and the amount and type of debris it contains. In addition, as mentioned above, the area is not well defined, much of the material is very small, and not all the material is floating on the surface.

To provide some perspective on the problem, NOAA calculated the approximate cost of a hypothetical cleanup. Considering an area that encompasses a significant part of what has been designated the Pacific Garbage Patch, about 385,000 square miles, it would take 67 ships, working 10 hours a day, covering swaths 650 feet wide, a year to accomplish it. Ship time alone would cost between $122 million and $489 million, without factoring in labor and equipment, and much of the plastic occurs in pieces too small to be scooped up with nets. To complicate matters further, the nets being used to strain out the plastic and other debris would also be collecting fish and other marine life. Attempting to clean up the global plastic patches really isn't realistic or practical and has many hidden costs and side effects. We need another approach.

## HOW TO FIX THE PROBLEM? WHERE DO WE GO FROM HERE?

While cleaning up beaches is helpful and beneficial, the ultimate solution to the problem of plastic and marine debris along the shorelines and in the oceans of the world is not cleanup and removal but prevention: cutting off or eliminating the plastic, Styrofoam, and other material at the source. It means taking every action necessary to keep the trash from getting into our rivers, waterways, and oceans to begin with.

Every discarded plastic bottle or bag, every cast-off fishing net, each piece of Styrofoam, each disposable diaper, six-pack ring, or cigarette lighter has a human being behind it. Someone dropped it, threw it away, out a window, or off a ship, and cast it adrift to end up in the ocean or on a beach. While some of these numbers, whether tons of plastic or billions of beverage bottles and shopping bags, were listed earlier in this chapter, their widespread presence along the shorelines of the world and in the identified garbage patches provide ample evidence of the scope and seriousness of the problem. Their longevity and lethal or sublethal effects on marine life and the economic costs should be evident to all of us but sadly are not. It comes down to production and advertising and then behavior and individual choices, what we buy and how we dispose of it.

Education and accessible information increase awareness, although this is often easier in some cultures and geographic areas than others. But it all has to start somewhere, and there are a number of conservation and environmental organizations actively involved in efforts to reduce plastic consumption and disposal. Save our Shores, Surfrider Foundation, Rise Above Plastics, Keep the Sea Plastic Free, and Bay Keeper are a few of these. Becoming aware of the plastic problem and the impacts of debris disposal helps us make better choices. While it makes sense to almost all of us to reduce the rate of production, consumption, and disposal of plastic, the big industries that produce plastic are not going to go away quietly. Constant attention, pressure, and legislation are necessary; aggressive incrementalism is a good approach.

Major flooding from a typhoon in Bangladesh in 1988 and 1989 was made more severe because plastic bags blocked flood drains. In response,

Bangladesh became the first country in the world to ban plastic bags, in 2002. In the same year, Ireland responded to the problem of plastic bags by implementing a heavy tax—15 cents per plastic bag—and reduced their usage by 90 percent in the first year. The tax was subsequently raised to 22 cents. These examples, as well as what seems to be an increasing global awareness of the many problems caused by plastic in the environment, have led to additional plastic bag bans.

Rwanda has a reputation for being one of the cleanest nations in Africa, indeed in the world, and banned plastic bags years ago. Italy, which historically has used 25 percent of all plastic bags in Europe, terminated single-use, nonbiodegradable plastic bags as of January 1, 2011. Bulgaria imposed a tax on plastic bags beginning on July 1, 2011. Yangon, Mexico City, Delhi, and Mumbai have all banned plastic bags, as have San Francisco, Los Angeles, Portland, Seattle, Austin, Chicago, and a number of other cities and counties in the United States.

The State of California banned plastic bags at checkout counters after a fierce legislative battle in 2014 that pitted the plastic bag industry against environmentalists. The ban was scheduled to go into effect in July 2015, but the plastic bag industry collected enough signatures on a referendum to get it on the November 2016 ballot to overturn the ban. This is discouraging, but we need to have faith that educated people with no economic interests will understand the problems we have surrounded ourselves with and make the right decision in the voting booth. In the 2016 election, California voters upheld the plastic bag ban. Use reduction and recycling, local ordinances banning the single use of plastic bags and single-use plastic water bottles, enforcement of the existing international bans on disposal of plastic and other marine debris waste from ships, and development of biodegradable materials are all parts of the solution.

# CHAPTER 8

# Petroleum and the Coastal Zone

Although everyone on the planet needs reliable sources of energy, more often than not it is in the coastal regions where our thirst for energy plays out and where the impacts of exploration, development, production, and transport of oil and gas are felt the most. We have become almost completely dependent on hydrocarbons, but it is time to recognize that they are finite and not renewable in our lifetimes. In addition, their extraction and transport comes with substantial impacts and risks, and their combustion over the past century has had a measurable and significant impact on our atmosphere and global climate. Yet we are now an energy-intensive society and economics or the actual price of energy, rather than external societal costs or sustainability, has continued to be the major driving force behind our energy choices.

Many of us understand that we need to reduce our dependency on fossil fuels, that the reserves remaining are limited, and that there are serious environmental costs from our reliance on them, yet sustainable renewable alternatives still only provide a small percentage of our total energy needs. How do our present practices and dependencies affect our coasts, and what role does or can the coastal zone play in our transition to sustainable energy sources?

## PETROLEUM: FORMATION AND EARLY USE

Petroleum is a product of the coastal ocean. Millions of years ago, trillions of diatoms and related marine phytoplankton (tiny floating plants) sacrificed their microscopic bodies to power our cars, trucks, motorcycles, airplanes, and ships—in short, most of our global industrial society. Massive blooms of plankton occur in the coastal and surface waters of the ocean when the right conditions converge. We can observe this at times when one particular type of plankton is abundant enough to actually turn the ocean a red, yellow, or brownish color. This usually happens in late spring but often in summer and into fall as well. As the hours of sunlight increase in the spring and when wind moves surface waters offshore and cooler, nutrient-rich bottom water rises to the surface in a process known as upwelling, the phytoplankton respond just like the weeds in your garden. Through photosynthesis, these tiny plants convert water, carbon dioxide, and the critical nutrients (nitrate, phosphate, and iron) into simple hydrocarbons.

While some of these algae enter the food chain and are consumed by zooplankton (e.g., krill) and also by small fish like sardines and anchovies, huge numbers of them sink to the seafloor, where they accumulate to form organic-rich muds. If the bottom waters are low in oxygen or if sediment deposition rates are high enough that the organics contained in the plankton are buried before they can be oxidized and decomposed, we can preserve these simple hydrocarbons. Over millions of years, as long as abundant plankton production continues to occur in the overlying waters, hundreds or even thousands of feet of organic-rich sediments can accumulate.

The burial and gradual subsidence of these thick layers of sediments can create a seafloor pressure cooker where increased pressure and temperature at depth slowly cook the preserved organic matter and convert it to petroleum: oil and natural gas (figure 8.1). While the right combination of conditions that create, preserve, and then convert organic matter to petroleum does not occur in seafloor sediments everywhere, they occurred frequently enough over the past 300 million years or so that there is a lot of oil and gas scattered around the planet. Originally a product of the coastal ocean or other productive marine environments, petroleum can now be found almost worldwide as tectonic plates have separated and collided, the seafloor has been uplifted to form mountain ranges, and old oceans have disappeared and new ones formed.

You may be surprised to learn that petroleum is actually the second most abundant liquid on Earth, after water, and where oil seeped to the surface naturally, ancient people found many uses for it. Six thousand years ago, and perhaps earlier, the tarlike substance was used to caulk or waterproof primitive boats, baths, and pottery or other vessels for

1. Marine plants and animals die and sink to the bottom of the seabed.

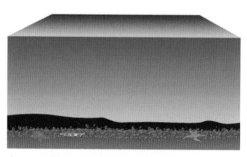

2. The plant and animal layer gets covered with mud.

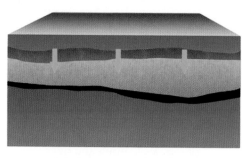

3. Over time, more sediment creates pressure, compressing the dead plants and animals into oil.

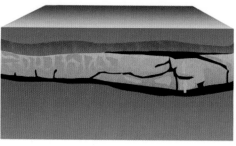

4. Oil moves up through porous rocks and eventually forms a reservoir.

FIGURE 8.1. Oil forms from organic material preserved in seafloor sediments that has been subjected to increased temperature and pressure.

holding water. It was also used as mortar to hold building stones together and even for paving streets in ancient Baghdad. Wells were dug for petroleum and gas in China by at least 347 C.E., and later in the Middle East. Distilling crude oil to produce kerosene-like fluids for lamps was a process that was first developed about a thousand years ago but was considerably improved in the mid-nineteenth century in Scotland, where through slow refining a number of useful products were distilled from petroleum, including paraffin, kerosene, and lubricating oils. Whale oil was originally used in lamps, but going out to sea for months at a time to hunt whales was a tough way to make a living and also a lot of work to fuel a lamp. When kerosene was first produced from oil it was found to burn cleaner and to be less expensive, and you didn't any longer have to go out searching for whales. The use of kerosene expanded quickly, and a large demand for petroleum soon followed.

Edwin Drake is usually recognized for drilling the first commercial oil well in the United States, near Titusville, Pennsylvania, in 1859. Wells had been dug earlier by hand, but Drake is often given credit because his well was actually drilled, using a steam engine, and because it led to the first oil boom. With the introduction of the internal combustion engine in the early twentieth century and then the first automobiles, demand exploded, and oil booms soon took place in Ohio, Oklahoma, Texas, and California. International discoveries were also being made early on, in Sumatra, Venezuela, and Mexico.

Although coal was the most common fuel used around the world until about the mid-twentieth century, the convenience of gasoline and other liquid fuels refined from petroleum for use in an expanding fleet of motor vehicles (motorcycles, cars, trucks, ships, and airplanes), as well as the increasing use of natural gas, took petroleum to the forefront of energy sources. Entering the second decade of the twenty-first century, about 90 percent of the fuel used by vehicles of all kinds is produced from oil, and in the United States oil meets about 40 percent of our total energy needs. In the 1950s the United States was exporting oil, but shortly thereafter we began to import oil to meet expanding domestic needs. By the 1970s we were satisfying our increasing demand by importing about two-thirds of our oil. As of 2016 the United States used about 20 million to 22 million barrels per day (one barrel = 42 gallons), and in recent years about 40 percent of that has been imported.

The estimation of oil reserves is an important but difficult business because of the geologic complexities beneath the Earth's surface. Although we have developed lots of sophisticated geophysical tools for imaging the subsurface, until you actually drill a well and start pumping oil, it is difficult to know for certain just what lies 10,000 or 15,000 feet beneath the surface. Terms like *accessible reserves, proven reserves, unproven reserves, technically and economically recoverable reserves,* and *estimated reserves* are all used to describe different levels of certainty with regard to oil deposits. The actual volumes of reserves listed for individual oil fields or different countries change quite regularly as the price of oil changes, as new technologies are developed, as exploration companies drill deeper, and as offshore exploration moves into deeper water farther offshore.

About 80 percent of the total global oil reserves are found in the Middle East, with nearly two-thirds (62.5 percent) beneath just five nations: Saudi Arabia, United Arab Emirates (UAE), Iraq, Qatar, and Kuwait. This area was a warm shallow sea 150 million to 200 million years ago, and conditions were ideal for lots of plankton production and preservation of their organic remains in the bottom sediments. Plate tectonics conspired to compress this ancient sea as Africa pushed into Europe and Asia, squeezing and folding the thick layers of sediments with their rich deposits of oil buried beneath the floor of this ocean basin. The waters of the sea gradually drained off to the east and west, but the compressed and folded oil-rich sediments became the mountains and deserts of Saudi Arabia, Kuwait, UAE, Iraq, Qatar, and their neighbors.

## OIL IN THE SEA

Massive oil tankers split open on coastal rocks or the dramatic blowout of drilling platforms, such as the *Deepwater Horizon* in the Gulf of Mexico in 2010, usually grab the attention of the news media and the public, simply because of the visual impact of oiled birds and beaches. Yet in a detailed study that evaluated sources of oil in the sea for a ten-year period (1990–99) by the U.S. National Academy of Sciences, and which has not yet been updated, the largest contributors to oil in the sea are neither tankers nor drilling platforms (figure 8.2). The midpoint of the estimates of the total amount of oil reaching the sea each year was about 1,430,000 tons, or 9,282,000 barrels; however, because of a whole variety of uncertainties and unknowns, this is a difficult value to determine on a global or even a regional scale, so estimates ranged widely from 3.4 million to 60 million barrels. It may come as a surprise that the largest percentage of that petroleum, whether global or in

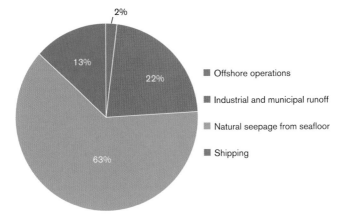

2%

13%

22%

63%

■ Offshore operations

■ Industrial and municipal runoff

■ Natural seepage from seafloor

■ Shipping

FIGURE 8.2. Oil can enter the ocean off of North American nations from a number of different sources.

North American waters, comes from natural seepage (46 and 63 percent, respectively), with the Santa Barbara Channel area of Southern California a big source. Oil consumption comes in at number two, making up 37 percent of the oil reaching the ocean globally and 35 percent in North American waters. This includes both land-based discharges from consumption and ocean-based operational discharges from ships. Oil from ship accidents, primarily oil tankers, makes up 12 percent of the global oil spillage and 4 percent of the oil reaching North American coastal waters. Drilling accidents and related releases are actually the smallest in total volume, 3 percent of global oil discharge and just 2 percent of North American releases, but often the most visible and disturbing.

DRILLING FOR OIL

The same geologic conditions that led to the formation of petroleum in the Middle East, but on smaller scales, have occurred in countless other coastal marine environments around the planet over the past 300 million or so years. Through the process of plate tectonics, as landmasses have broken apart and collided, as new ocean basins have formed and old ones have been consumed, and as ocean floor sediments have become accreted onto continents as sedimentary rocks, oil companies have used constantly evolving technologies to explore, probe, and image the subsurface looking for sedimentary deposits that may contain oil.

Today, as many of the easily discoverable fields on land have been located and exploited, efforts are being focused in offshore areas, and now, somewhat cautiously, in the Arctic, and deeper beneath the continents, in rocks that do not give up oil easily and thus are being exploited by means of hydraulic fracturing ("fracking").

In the 1920s and 1930s oil began to be pumped from beneath the marshes and bayous of Louisiana, often from wooden platforms built on piles of old oyster shells. The first true offshore platform that produced petroleum was constructed in 1937 about a mile offshore from Louisiana in just 14 feet of water. As oil demands grew following World War II, surplus military barges were often used as drilling platforms and were gradually moved farther offshore into deeper water. Ownership rights soon became a controversial issue, however, which was resolved in 1953 when the Submerged Lands Act gave states the rights to most natural resources within 3 miles of their coastlines. The federal government was able to auction off lease blocks of submerged land farther offshore. Because of the vast oil deposits beneath the Gulf of Mexico, by that time there were already about seventy offshore drilling and production platforms in water out to a depth of 70 feet, almost the shoreline by today's offshore standards.

As demands for oil and its refined products continued to soar, and with the rise in oil prices following the formation of the Organization of Petroleum Exporting Countries (OPEC) in 1960 and the 1973 oil embargo, petroleum development progressed into deeper and deeper water in the Gulf of Mexico (figure 8.3). Drilling in 1,000 feet of water was accomplished by 1973, and in 1987 floating platforms were drilling in water a mile deep. This trend continued, with hundreds of wells drilled, completed, and now pumping oil from deep water. As of early 2016, there were fifty-eight active drilling platforms operating in the Gulf in water depths ranging from 660 feet to nearly 2 miles, with names like *Noble Globetrotter, Rowan Reliance, T.O. Deepwater Invictus, and Diamond Ocean Blackhornet.*

On April 20, 2010, there was another, better-known platform working in the Gulf, the *Deepwater Horizon,* a nine-year-old semisubmersible drilling rig that was boring a deep exploratory well 18,360 feet below sea level in water close to a mile deep, 40 miles offshore, when chaos, death, and destruction ensued. There are serious challenges in drilling in very deep water, not the least of which is you can't really see what's going on down there, and even if you could, it's not very easy to

FIGURE 8.3. The P-51 platform drilling off the coast of Brazil in 2009 will produce about 180,000 barrels of oil and about 7.8 million cubic yards of gas per day when in full operation. (Photo: Divulgação Petrobras/ABr [Agência Brasil] licensed under CC BY 3.0 via Wikimedia Commons)

do much about it if there's a problem. This accident is discussed in more detail below.

On the western edge of the United States oil was first discovered in California's Central Valley in 1865, near Santa Barbara in 1886, and then in the Los Angeles area in 1892. The oil boom changed Los Angeles forever, as more and more people flooded in, hoping to get a share of the action, not unlike the California gold rush forty-three years earlier. In 1921, a drilling rig on Signal Hill in the Long Beach area hit a gusher, and this soon became the state's largest oil field. California quickly became the largest oil-producing state in the nation, and in 1923, 25 percent of the world's entire production of crude oil came from the coastal area of Southern California.

A hundred miles northwest of Signal Hill, in the small beach community of Summerland, on the coast between Ventura and Santa Barbara, the natural oil and asphalt seeps along the shoreline had attracted attention and prospectors for some time. By 1895 the search for black gold was going full bore, with wooden drilling derricks being erected along the bluffs and beaches of the formerly quiet spiritualist community.

FIGURE 8.4. Oil piers and wells along the shoreline of Summerland, Santa Barbara County, California, in the early 1900s. (Photo: G.H. Eldridge, via Wikimedia Commons)

The oily sheen on the surface of the ocean just offshore and the globs of tar seen on the beaches hinted to the early oil prospectors that there must be oil offshore as well. So in 1886, 130 years ago, the first piers were built extending out from the beach in order to drill into the shallow offshore seafloor. This was nearly a century before the California Coastal Commission and before environmental impact reports, however, and soon an array of piers with dozens of individual wells had been drilled and the oil started flowing (figure 8.4).

The next big step for the embryonic offshore oil industry took place just off Seal Beach, near Long Beach, where the first artificial island was constructed in 1954 in order to provide a solid platform for reaching the tentacles of the drill bits a little farther offshore. Things were now starting to get exciting for the oil industry. Exploratory drilling offshore from Summerland struck oil in the same geologic formation that had produced the black gold along the shoreline but in water 100 feet deep. In 1958 the first offshore drilling and production platform in California, named *Hazel*, was towed to the site on a barge, lowered to the seafloor, and anchored in what became the Summerland Offshore Oil Field. Two years later, *Hazel* was followed by *Platform Hilda*, and then to round out what became known as the 4-H platforms, *Hope* and *Heidi* were installed in 1965, all in state waters within 3 miles of the coastline.

Over the next thirty years, before they were decommissioned in 1996, these four platforms would pump out a total of 62 million barrels of oil and 131 billion cubic feet of natural gas. At the present rates of oil consumption in the United States, about 22 million barrels a day, all of the oil pumped out over the roughly thirty-year lifetime of the 4-H platforms would not last three days.

California's offshore drilling soon moved into federal waters and into high gear. By 2012 there was an alphabet soup of nine active drilling and production platforms in state and municipal waters and twenty-three active platforms in federal waters spread out between Huntington Beach on the south and Point Arguello on the north. The area between Seal Beach and Huntington Beach hosted the offshore platforms *Esther, Eva, Emmy, Edith, Ellen*, and *Eureka*. The area farther west along the coastline near Ventura was graced with platforms *Gina, Gail, Gilda*, and *Grace*. And the Santa Barbara Channel harbored the platforms *Heritage, Harmony, Hondo, Habitat, Hillhouse, Henry, Hogan*, and *Houchin* and a few more near Point Conception: *Hermosa, Harvest*, and *Hidalgo*. Offshore oil drilling and production was a big business.

## HAZARDS OF OFFSHORE DRILLING

By the very nature of the offshore drilling operation, there are significant hazards. High wave and wind conditions, earthquakes, icebergs in high latitudes, and slope stability and unstable bottom conditions all contribute to an often-hazardous environment. Working in increasingly deeper water and boring deep into the seafloor, where volatile and flammable oil and gas are under very high pressure, involve a substantial degree of uncertainty and risk, but the financial rewards can be huge. In 2006 the Gulf of Mexico alone had 3,858 oil and gas platforms. In the decade between 2001 and 2010, however, the U.S. Minerals Management Service reported 858 fires and explosions on offshore rigs in the Gulf of Mexico, which led to 1,349 injuries and 69 deaths. Occidental Petroleum's *Piper Alpha* offshore production platform in the United Kingdom's sector of the North Sea exploded after a gas leak on July 6, 1988, causing the death of 167 workers.

With literally thousands of drilling and production platforms now in place off the coastlines of dozens of countries around the world, the hazards, errors, shortcuts, and cost savings that have led to major spills, accidents, and blowouts have often been highly publicized. We ought to learn from past disasters and not repeat the same mistakes. The offshore drilling industry, however, is a complex undertaking that involves many different agencies and entities. Leasing regulations, operator history, permits, and precautions vary by nation, and even under the best of conditions, or sometimes the worst, accidents can still happen, with far-reaching and long-lasting consequences. While one can argue that the overall history of offshore drilling has been a relatively good one, there are a number of well-publicized disasters that could have been avoided had appropriate industry standards been followed, necessary precautions been taken, and inspections and government oversight been more stringent. Two examples below illustrate what can go wrong and why.

## The 1969 Santa Barbara Oil Spill

The blowout of Union Oil's *Platform Alpha* in the Santa Barbara Channel on the morning of January 28, 1969, was the first major event of its kind off the West Coast, and it has often been considered a catalyst for the beginning of the environmental movement in California. The platform was being operated by Union Oil but in partnership with Gulf, Texaco, and Mobil and was drilling in 275 feet of water. The Santa Barbara Channel is sliced up by a number of faults, and natural oil seepage had been recognized for centuries. On this morning the consortium was finishing the fifth of the fifty-five wells ultimately planned for *Platform Alpha*. While the well was completed to a depth of 3,500 feet, in order to save money the well casing (the steel pipe inserted into the drill hole to prevent collapse and to prevent any oil from escaping) was only installed to a depth of 245 feet into the seafloor. The drilling had actually gone through two fault zones in its 3,500-foot depth. As the drill string was being extracted from the hole, it was pulling the high-density drilling mud up the hole with it. That led to the accumulation of high pressure in the well beneath the drill bit, which pushed the drill pipe upward onto the deck of the platform. This ruptured the hose that pumped the drilling mud into the well, and the natural gas under high pressure then blew out the rest of the drilling mud, split the well casing, and then broke through on the seafloor beneath the platform. Oil and gas soon emerged on the ocean surface around the platform. Full blowout occurred, and the 1969 Santa Barbara oil spill was now in process.

The blowout was due in large part to the request to use a shallower casing than federal standards required but was approved by the U.S.

FIGURE 8.5. The 1969 Santa Barbara oil spill cleanup operation was a long, messy ordeal that used primitive methods and ended up killing a large amount of intertidal life. (Photo: Bob Duncan © 1969)

Geological Survey. Oil and gas erupted on the seafloor for eleven days while several attempts were made to plug the hole and stem the spill. Ultimately, the seafloor cracks were sealed with a chemical grout but not until 70,000 barrels of oil (3 million gallons, or enough to fill four and a half Olympic-size swimming pools) had spread out over 800 square miles of ocean surface and coated 34 miles of shoreline, shoreline that was home to people who had opposed the oil drilling to begin with. Impacts on wildlife were large; oil-covered beaches, the intertidal zone, and boats were highly publicized, and the industry got a black eye that stuck around for years (figure 8.5). The industry was not prepared for a blowout, nor was it prepared or willing to clean up afterward.

## The 2010 BP Deepwater Horizon Disaster

On April 20, 2010, forty-one years after the Santa Barbara blowout and spill, the British Petroleum *Deepwater Horizon* semisubmersible drilling rig, working 40 miles off the coast of Louisiana in water nearly a mile deep on the Macondo well, suffered a catastrophic series of explosions. A supply boat, tied up to the rig, which was offloading used

FIGURE 8.6. Platform supply vessels battle the blazing remnants of the offshore drilling rig *Deepwater Horizon* in the Gulf of Mexico, after explosion. (Photo: U.S. Geological Survey)

drilling mud, witnessed the first stage of the disaster. Partway through their routine but nighttime operation, the crew suddenly noticed an eruption of mud and seawater, twenty stories above the main deck of the platform, blasted upward by a flood of natural gas from 3 1/2 miles below the level of the rig. Within seconds there was a huge explosion and fireball, which plunged the entire rig into darkness and showered the supply boat with mud and debris. Then came a second and larger blast, and a third, as the captain watched some of the BP *Deepwater Horizon* crew members jumping from the platform into the dark waters of the Gulf of Mexico 65 feet below. The massive rig, the size of a football stadium, burned for two days (figure 8.6), gradually tilting over until it passed the point of no return. It broke off the mile-long steel pipe that connected it to the wellhead on the seafloor below and sank to the bottom of the Gulf, and then the oil started coming. It continued gushing for eighty-seven days and was finally capped on July 15, after an estimated 4.9 million barrels of oil (210 million gallons, or enough to fill 318 Olympic-size swimming pools) had entered the Gulf (figure 8.7).

FIGURE 8.7. Oil from the *Deepwater Horizon* spill approaching the coast of Mobile, Alabama, May 6, 2010 (Photo: Michael B. Watkins, Petty Officer 1st Class United States Navy).

Oil continued leaking, however, and was not declared totally sealed off until two more months had passed, on September 19, 2010. Almost two and a half years later, in January 2013, oil slicks were reported at the surface that matched the Macondo well.

Eleven crew members died in the accident, and the blowout released the largest accidental marine spill in oil company history. The disaster generated a large response in an attempt to protect estuaries, wetlands, and beaches around the Gulf. Every method in the arsenal of cleanup technology was used, with varying degrees of success: floating booms to keep oil away from wetlands, skimmer ships to suck up floating oil, controlled burns of oil at the sea surface, spraying 1.84 million gallons of oil spill dispersant, and shoveling oil-contaminated beach sand into plastic bags—high technology indeed. Although new oil spill cleanup methods always seem to operate under ideal conditions in calm water, when the chips are down and the ocean gets riled up all bets are off, and cleanup operations usually turn to scooping up oil on beaches with hand tools. The effects on a wide range of marine life, including birds, turtles, dolphins, and fish were well documented, as was extensive damage and losses to marine and wildlife habitats, as well as the fishing and tourism economies.

Complicating any offshore drilling accident is the typical complexity of responsibility due to the large number of parties involved. The *Deepwater Horizon* was built by a South Korean company, owned by Transocean, operated under the Marshallese flag of convenience, and chartered to British Petroleum (BP). Although BP was the operator and principal developer of this lease prospect with a 65 percent share, 25 percent was owned by Andarko Petroleum Corporation and 10 percent by MOEX Offshore 2007, a unit of Mitsui.

It is quite difficult to destroy and sink a huge offshore drilling rig, which in many ways was really a floating palace. It could accommodate a crew of 160, had a movie theater, a gym and sauna, in-room Internet and satellite TV, a heliport, and a dynamic positioning system such that it remained very stable even in rough weather.

But drilling in a mile of water, where virtually everything is done remotely in a hostile environment—things on the seafloor are very cold, extremely dark, and under very high pressure—is a complicated undertaking. Remote vehicles did the work where humans couldn't go, drilling, welding, and fitting pipe together. While the technology has continued to advance, both in exploring for oil with seismic imaging of the deep-sea floor and in drilling rigs and undersea vehicles for drilling and completing wells, operating in deeper water and at great depths beneath the seafloor presents new problems and challenges but also opportunities for huge profits.

The Macondo blowout and spill became one of the most intensely studied marine disasters in U.S. history, with investigations, commissions, and committees appointed by a variety of agencies and organizations tasked to determine what happened and why. When all was said and done, however, there was agreement that a chain of decisions and events, large and small, individual and corporate, had conspired to produce the BP *Deepwater Horizon* blowout that should never be allowed to occur again anywhere. A detailed investigation by *Fortune* magazine concluded that the disaster was a long time in the making, the product of a corporate culture that dignified risk taking, despite a repeated commitment to safety. While various groups have produced many safety standards over the years, many of them were irrelevant. British Petroleum needed to do a better job bridging the disconnect between management and the people on the drilling rig. In hindsight, there was nothing about the *Deepwater Horizon*'s drilling of Macondo No. 1 that had gone well. Prior to the blowout, there had been a number of significant problems, stuck drills, and lost circulation of drilling mud; all were time consuming and dangerous such that the operation had fallen behind schedule and was $58 million over budget. Some of the drilling crew began calling it "the well from hell," and things got worse. The various reports cited a number of serious shortcomings and cost-cutting decisions—defective cement in the well, inadequate sealing, and insufficient safety systems—and concluded that the spill resulted from "systemic" root causes and "absent significant reforms in both industry practices and government policies, might well recur."

Legal proceedings were resolved with British Petroleum pleading guilty to eleven counts of manslaughter, two misdemeanors, and a felony count of lying to Congress. BP also agreed to four years of government monitoring of its safety practices and ethics and was temporarily banned from new contracts with the U.S. government. In July 2015 BP agreed to pay $18.7 billion in fines, the largest corporate settlement in U.S. history. All criminal and civil settlements as well as payments to a trust fund reached a total of $42.2 billion, and BP was ruled the responsible party because of its gross negligence and reckless conduct.

The *Deepwater Horizon* blowout and subsequent disaster led to the largest corporate fines and penalties in U.S. history. It also occurred in waters that were thought to be as tightly controlled and regulated as those anywhere in the world. Yet it happened with far-reaching and long-lasting results, one of which was the complete restructuring of the government regulatory system for offshore drilling. Regulatory decisions made by the Minerals Management Service (MMS) that contributed to the 2010 spill included (1) the decision that an acoustically controlled shut-off valve would not be required as a last resort against underwater spills at the site; (2) the agency's failure to suggest other failsafe mechanisms after a 2004 report raised questions about the reliability of the electrical remote-controlled shut-off device; and (3) the agency's granting of permission to British Petroleum and dozens of other oil companies to drill in the Gulf of Mexico without first getting required permits from NOAA that would have assessed threats to endangered species and to assess the impact of drilling on the Gulf. In May 2010, Secretary of the Interior Ken Salazar dissolved MMS and replaced it with the Bureau of Ocean Energy Management (BOEM), the Bureau of Safety and Environmental Enforcement (BSEE), and the Office of

Natural Resource Revenue, which would be organized to separately oversee energy leasing and environmental assessment, safety enforcement, and revenue collection, respectively.

Virtually any offshore area in the world could experience an accident similar to the Macondo well blowout. Leasing and drilling in deepwater conditions will always come with significant potential risks, whether understood and planned for or not. When potential corporate profits are high and the lease sales and government income is significant, opportunities for less than thorough analysis and precautions will always be present. The mistakes, shortcuts, and cost-cutting practices of the past and the environmental impacts that have resulted should give pause and reason for more careful scrutiny and continuous control and monitoring of all future offshore drilling activity.

## THE GLOBAL TRANSPORT OF OIL

Globally about 12 percent of all the oil entering the ocean comes from intentional or accidental release from oil tankers. Most of the major sources and suppliers of oil are in different places on the planet from the major consumers, so there is a very large volume moving around somewhere in the global ocean in large tankers, leaving one set of ports and heading for others (figure 8.8). Volumes of oil exports and imports change somewhat from year to year, due in large part to the price per barrel, but in recent years they range from about 45 million to 50 million barrels a day, or enough oil to fill about 3,000 Olympic-size swimming pools. The major exporting and importing regions do remain reasonably consistent, however, with most of the planet's oil being exported from the Middle East (~40 percent), Russia and the former Soviet Union nations (~19 percent), Nigeria and North Africa (~14 percent), and Central and South America (~10 percent). The five largest oil-exporting nations over the past twenty-five years have been Saudi Arabia, Russia, Iran, UAE, and Nigeria. Most of the oil is transported in supertankers to places not particularly close to the sources, with about 41 percent going to Asian nations, 26 percent going to the European Union nations, and about 18 percent to the United States, the single largest global importer of oil, about 8 million to 9 million barrels a day. After the United States, Japan, China, India, and South Korea have been the largest users of imported oil. In 2011 the world consumed oil at a rate of 93 million barrels per day, and over half of all that oil was imported from somewhere else.

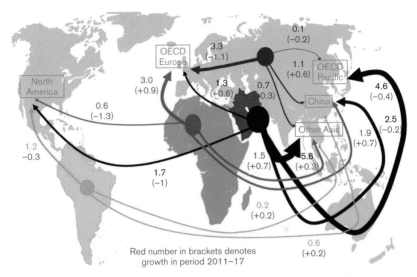

FIGURE 8.8. Supertankers move millions of barrels of oil around the world's oceans from oil-rich nations to those who need to import oil. All values are in millions of barrels per day.

As world appetite for oil increased, the volumes moved by sea also increased, along with the size of the tankers transporting it. During World War II typical tankers were in the 16,000-ton range, carrying about 114,000 barrels. Ships grew gradually in size to over 500,000 tons and about 1,300 feet long in the 1980s. The longest tanker ever built, the *Seawise Giant,* stretches to 1,504 feet; it was in service from 1979 to 2009 and then converted to an oil storage terminal. For comparison, the Empire State Building in New York City is 1,472 feet high. Ships this large usually extend 78 to 85 feet below the waterline, which severely limits the ports they can enter and also the places they can go when troubles arise. The largest tankers in operation at present are 1,247 feet long and can carry about 3 million barrels of oil, or nearly 133 million gallons (figure 8.9). This is forty-five times more oil than entered the ocean during the 1969 Santa Barbara oil spill and 61 percent of the total oil released during the BP *Deepwater Horizon* disaster in 2010, moving around in one big tank. One of these supertankers (at this size they are called ULCCs, for Ultra Large Crude Carriers) could carry over one-third of the volume imported by the United States on a typical day.

Growth in tanker size has been driven primarily by economic arguments. By reducing operating costs and increasing cargo capacity, ship

FIGURE 8.9. The MT *Hellespont Alhambra* on her maiden voyage in 2002, carrying nearly 3 million barrels of crude oil. This is one of the four largest ships in the world by gross tonnage. (Photo: PA2 Dan Tremper, USCG, via Wikimedia Commons)

owners foresaw great savings. The increase in tanker accidents and the subsequent lawsuits and cleanup costs, however, raised insurance rates for the large ships. Profit margins also declined with lower crude oil prices. Structural problems and tank explosions in some of the early supertankers raised the investments as well. These 1,200- to 1,300-foot-long tankers also have limited maneuverability because of their huge size and mass, making it very difficult to respond in an emergency. It takes about 20 minutes and 4 miles for a 250,000-ton tanker cruising at 16 knots (17.6 mph) to stop or avoid a collision. Guiding and steering a ULCC from a bridge nearly a quarter of a mile from the bow, 100 feet above the water, and 150 to 200 feet from one side to the other, also present challenges that did not exist with smaller ships. In addition, most older navigation charts were prepared for much shallower-draft vessels, so they may not be accurate at the depths these ships need.

With improvements in navigation, safety measures, and, in many cases, the construction of double-hulled tankers (which were required for vessels entering all U.S. ports following the 1989 *Exxon Valdez* spill), the total number of oil spills and also the volumes of oil released have continued to decline since about 1980, although it often is a single large accident that is responsible for much of the volume of oil that enters the ocean in any one year.

Many of the major accidents or groundings are known by the tanker names. There are some that are not necessarily the largest but happened to impact coastlines that were particularly sensitive, populated, or otherwise noticed. The *Torrey Canyon,* the *Amoco Cadiz,* and the *Exxon Valdez* were all incidents that received global attention, and for different reasons, but are good examples of what can go terribly wrong very quickly with a large oil tanker.

*The* Torrey Canyon.

The *Torrey Canyon* spill resulted from one of the earliest and largest tanker groundings and exemplified the international nature of the oil transport enterprise. The ship was carrying Saudi Arabian crude oil, was owned by one American oil company but leased to another, was registered in Liberia, had an Italian crew and a German captain, and was delivering oil to Milford Haven in West Wales. Who was responsible? It ran aground in the spring of 1967 on the southwest coast of England in broad daylight on a well-marked reef. While not large compared to present-day supertankers, it still spilled 850,000 barrels (35 million gallons) of crude oil. To provide some idea of how much oil this is, take a football field and build a wall around it 100 feet high and fill it with oil. That's about 35 million gallons. At the time the world's largest spill, nearly fifty years later it remains the United Kingdom's worst. Ultimately the oil covered hundreds of miles of coastline on both sides of the English Channel, affecting the United Kingdom, France, and Spain.

The accident resulted from a series of mechanical and human errors that multiplied the risk factors. The tanker did not have a scheduled route, so it lacked all the necessary navigation charts for the Scilly Islands off Cornwall. The vessel was using LORAN (Long Range Navigation), as this was before routine satellite navigation, rather than a more accurate Decca Navigator system. As a collision with a fleet of fishing vessels threatened, confusion ensued between the captain and the pilot as to their exact position and also whether they were in fact in manual or automatic steering mode. When these questions were finally answered it was too late to avoid grounding, and the *Torrey Canyon* ended up on the Seven Stones Reef (figure 8.10).

FIGURE 8.10. The 1967 *Torrey Canyon* breakup and spill off the coast of the United Kingdom was the first very large marine oil spill. (Photo: Courtesy of Tony Wheatcroft)

There had been very little experience with oil spills of this magnitude at the time, so dealing with it was a process of trial and error. When refloating the ship proved impossible and it began to break up, a military approach that involved dropping a combination of bombs, aviation fuel, and liquefied petroleum jelly was employed on the site in an effort to burn the oil, which proved ineffective as large waves extinguished the flames. Over 10,000 tons of "detergents," which were believed by many to be industrial solvents, were used in an effort to disperse the crude oil along the coastline. In the end, 48 miles of French coast and 115 miles of Cornish coast were covered in oil from a slick encompassing 270 square miles, ultimately killing about 15,000 seabirds as well as many miles of intertidal life. The solvents and other chemicals used were generally believed to be far more damaging to marine life than the crude oil itself, and the British government was heavily criticized for the way it handled the incident. Claims against the owners of the vessel and the subsequent settlement were the largest in history at the time for any oil spill and led to a number of changes in international shipping regulations and liabilities.

## *The* Amoco Cadiz

Eleven years later, in March 1978, a similar but larger disaster occurred, also in the English Channel, a major thoroughfare for petroleum moving from the Middle East to ports in northern Europe. The *Amoco Cadiz* was carrying 1.6 million barrels of crude oil from Saudi Arabia and Iran, destined for Rotterdam after a port call in Lyme Bay, United Kingdom. The ship was classed as a Very Large Crude Carrier (VLCC), sailing under a Liberian flag of convenience, and owned by Amoco, although the cargo of crude belonged to Shell Oil.

The ship met with gale conditions and heavy seas while in the English Channel. One large wave led to failure of the ship's rudder and steering gear, which had been built in Spain. While there were calls for a tug, which responded and was finally able to get a tow line on the tanker, due to the immense mass of the ship and the very strong winds the ship ran aground on the coast of France within twelve hours of the original distress call. Twelve hours later the ship broke in half, and the entire cargo of 1.6 million barrels leaked into the stormy Atlantic, nearly twice the volume as the disastrous *Torrey Canyon* spill.

Winds pushed the crude onto the coast of Brittany, where 43 miles of shoreline were quickly covered, but continued winds from the west moved the oil eastward to ultimately coat 192 miles of coastline. Tourist beaches, boats, and harbors were all covered with what was described as "chocolate mousse," an emulsification of oil and water, which complicated cleanup efforts. The loss of marine life at the time was greater than from any previous oil spill. Nearly 20,000 seabirds died, as well as tons of oysters and small crustaceans and echinoderms along miles of shoreline. Ultimately, Amoco, an American oil company, paid $120 million in damages to France ($440 million in 2015 dollars).

## *The* Exxon Valdez

Exactly eleven years later, the worst oil spill in the United States until the *Deepwater Horizon* blowout in 2010 took place when the captain of the *Exxon Valdez*, who had a history of alcohol violations, went below deck as the tanker, full of North Slope crude oil, was leaving Valdez, Alaska. Twenty-five miles from port, and using a modern navigation system, the officer on the bridge noted icebergs from nearby Columbia Glacier within the shipping channel. While maneuvering to avoid the icebergs, the tanker hit submerged rocks. The Coast Guard commander called it "almost unbelievable" that the *Exxon Valdez* had strayed from a 10-mile-wide shipping channel to crash into Bligh Reef. "This was not a treacherous area," he remarked. "Your children could drive a tanker through it." This accident and its aftermath led to an interesting quote: "Quite frankly, a supertanker leaving Valdez, Alaska, is no more dangerous than a mountain biker with a backpack containing 50,000,000 gallons of oil."

At 987 feet long, the *Exxon Valdez* was not a supertanker, but it was carrying about 55 million gallons, or 1.3 million barrels, of oil. Fortunately, only 22 percent of the cargo, 11 million gallons (262,000 barrels), spilled, and the remaining 42 million gallons were pumped out and salvaged. The oil spread over 1,100 miles from the collision, equivalent to the entire coastline of California, and ultimately blackened 3,000 miles of shoreline (figure 8.11). The entire region is a rich habitat for salmon, sea otters, seals, and seabirds. Best estimates are that from 100,000 to as many as 250,000 seabirds, 247 bald eagles, at least 2,800 sea otters, 300 harbor seals, 22 orcas, and an unknown number of salmon and other fish died as a result of the oil spill. Thousands of pages were written about the spill, its environmental impacts and the diverse cleanup methods used, their effectiveness or lack thereof, their public health hazards, and how long oil has persisted in the area.

As with any oil spill disaster, it is important to learn from the *Exxon Valdez* event and put regulations, requirements, and processes in place to reduce the likelihood of a spill like this happening again and to be prepared to respond if it does. Unfortunately, every new disaster seems to involve a different environment and a different set of conditions. With the *Exxon Valdez,* there were a number of factors that came into play:

- Exxon Shipping Company failed to supervise the ship's master and provide a rested and sufficient crew for the ship. The National Transportation Safety Board concluded that this practice was widespread throughout the industry, which prompted new safety recommendations.
- The officer on the bridge failed to properly steer the vessel, possibly due to fatigue or excessive workload.
- The ship's radar had been broken and disabled for more than a year before the disaster.

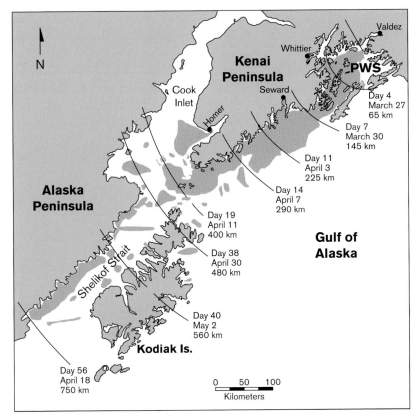

FIGURE 8.11. Known extent of the area covered by the 262,000 barrels of the 1989 *Exxon Valdez* Alaskan oil spill. The line in upper right points to site of tanker grounding, PWS—Prince William Sound. (Map: Mark Carls, NOAA/NMFS)

- The oil industry had promised but never installed state-of-the-art iceberg monitoring equipment.
- Lack of available cleanup equipment and personnel slowed the spill cleanup efforts.

Legal proceedings went on for nearly twenty years, with judgments against ExxonMobil of $4.5 billion. Through a series of ExxonMobil legal appeals and subsequent legal actions and decisions, which went all the way to the U.S. Supreme Court, this amount was finally reduced to $507.5 million in punitive damages, including lawsuit costs, about 11 percent of the original judgment. Exxon also spent an estimated

$2 billion cleaning up the spill and a further $1 billion settling related civil and criminal charges. Of the original 22,000 plaintiffs involved in the lawsuit against the biggest oil corporation in the world, during the twenty years of court battles, 6,000 had died.

The history of the spill and its causes, as well as the subsequent cleanup operations, raised many questions about tanker safety and management, the wisdom of cleaning up oil from shorelines, and uncertainty as to whether trying to clean up birds and otters is a productive and effective response. There are also some depressing legal lessons, among them that corporate lawyers can tie up any environmental damage settlement in court for decades, leaving plaintiffs holding the empty bag.

## WHERE DO WE GO FROM HERE? LOOKING TO THE FUTURE

Oil has been the world's leading source of energy since the mid-1950s because of its high energy density, ease of transport, and relative overall abundance. Being able to refine oil into a large number of fuels, ranging from diesel to gasoline and jet fuel, has allowed oil and its refined products to provide 32.2 percent of all global energy needs, and the world's people now consume about 93 million barrels of oil every day. In the United States we use about 20 million to 22 million barrels of oil a day, which provide about 36 percent of all our energy needs.

Nearly every aspect of oil intersects or interacts with the coastal zone. The formation of oil in offshore marine basins; the exploration, drilling, and production of the oil; and finally the global transport in and out of ports around the world all involve the coastline and coastal waters. As the easily accessible oil fields have been exploited, exploration and drilling have moved into deeper waters farther offshore and into the Arctic. These practices come with very significant challenges and risks. Oil has gotten into the sea from a number of sources, some natural but many from human mistakes or negligence. The environmental and legal costs of these accidents or spills have been very large in most recent cases, and the oil companies have made some critical mistakes and paid for them. Growing concerns about the potential impacts of additional offshore drilling led to the recent cancellation of plans to open up the Atlantic coast off the southern United States for leasing. The California coast similarly is not planning on additional leasing, and the Arctic is in question.

While modern society is still highly dependent on oil, the resources are finite, and we have reached the end of cheap or easily accessible oil. Burning oil, like any fossil fuel, produces large volumes of carbon dioxide, adding to our climate change and sea-level rise concerns. These are very large and far-reaching issues (see chapters 5 and 13) that we are going to be dealing with for decades to come. The most important step we can take as a global civilization is to do everything humanly feasible to transition to renewable sources of energy as quickly as possible in order to lower our dependence on oil and reduce our climate change footprint. We already know that this is not going to be easy, and there are going to be many debates and political battles ahead. But the tide is beginning to turn, and renewable energy sources are the fastest growing sector of the energy economy. This is good news.

CHAPTER 9

# Coastal Power Plants

## POPULATION AND ENERGY USE

Our modern global society consumes a lot of energy, and the amount we use continues to increase year after year. It is difficult to make comparisons of energy consumption between countries because the sources of energy used vary from region to region. Fortunately, someone has done the calculations and reduced energy use to one easy to understand and comparable unit: barrels of oil equivalent (BOE) per capita (or person) per year. So if we convert all of our energy sources, oil, gas, coal, nuclear, and hydroelectric and other renewables, into a heat equivalent, we can then use a barrel of oil as the equalizer.

Not surprisingly, the per capita use of energy around the world varies widely (figure 9.1), with the oil-producing states of the Middle East (Qatar, Kuwait, and UAE, for example) being at the top of the list, along with Iceland and Luxembourg. Canada and the United States are next on the list, followed by a number of European nations. The Republic of Korea and Japan fall into this same range, while China, Brazil, and India are significantly lower on the list, as are most of the nations of Africa, Central and South America, and South Asia. While per capita energy use in China and India and these latter areas are all increasing, the per capita use of energy for the United States, Japan, and the European countries are all slowly declining, which is a good sign for the planet. North America as a region uses the most energy per capita, followed by the

FIGURE 9.1. Per capita use of energy varies widely by country. (Courtesy *Burn and Energy Journal*. Sources: U.S. Energy Information Administration, International Energy Agency, CIA World Factbook, U.N. Dept. of Economics and Social Affairs)

ANNUAL CONSUMPTION PER PERSON
Millions of British thermal units (Btu)

- More than 400
- 250 to 400
- 150 to 249
- 75 to 149
- 25 to 74
- 10 to 24
- 5 to 9
- <5

1 million Btu = roughly 8 gallons of gasoline

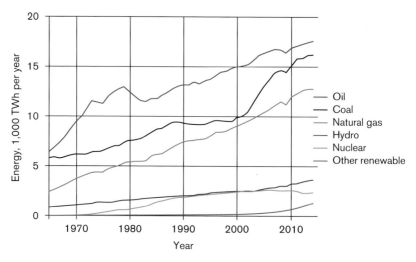

FIGURE 9.2. Fossils fuels still provide the great majority of the world's energy, and their use is increasing.

European Union, which uses less than half of North America. East Asia and the Pacific use significantly less energy per capita.

Looking at total energy consumption (population multiplied by per capita use), China now uses about 36 percent more energy than the United States because the population of China is 4.3 times larger. After China and the United States, the next highest consumers of total energy in order are India, Russia, Japan, Germany, Brazil, South Korea, Canada, and France.

Where is all of this energy coming from? On a global scale, as of 2014, as precisely as it is possible to determine, oil provided 32.2 percent of all energy used; coal, 30.0 percent; natural gas, 23.3 percent; and nuclear, 4.4 percent; all renewables generated the remaining 10.1 percent (figures 9.2 and 9.3a). Use of all fossil fuels continues to increase, while nuclear power is declining. The renewables were led by hydroelectric power at 6.7 percent, followed by wind (1.1 percent), biofuels (0.5 percent), solar (0.2 percent), and others (0.9 percent). The importance of renewables is increasing, although still at a relatively modest rate. We are still dependent on fossil fuels, which are nonrenewable, for 85 percent of all our energy needs. This is neither a sustainable nor a good long-term strategy, because the resources are clearly finite and their use causes well-recognized environmental impacts. Our total global energy consumption in 2014 was equivalent (BOE) to 92.8 billion barrels of oil, or 254 million barrels every day.

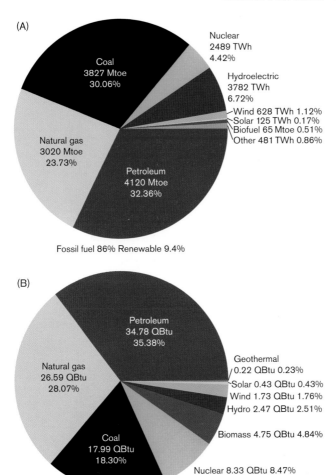

(A)

Coal
3827 Mtoe
30.06%

Nuclear
2489 TWh
4.42%

Hydroelectric
3782 TWh
6.72%

Wind 628 TWh 1.12%
Solar 125 TWh 0.17%
Biofuel 65 Mtoe 0.51%
Other 481 TWh 0.86%

Natural gas
3020 Mtoe
23.73%

Petroleum
4120 Mtoe
32.36%

Fossil fuel 86% Renewable 9.4%

(B)

Petroleum
34.78 QBtu
35.38%

Natural gas
26.59 QBtu
28.07%

Geothermal
0.22 QBtu 0.23%
Solar 0.43 QBtu 0.43%
Wind 1.73 QBtu 1.76%
Hydro 2.47 QBtu 2.51%

Coal
17.99 QBtu
18.30%

Biomass 4.75 QBtu 4.84%

Nuclear 8.33 QBtu 8.47%

FIGURE 9.3. (a) Global sources of energy (2013); (b) sources of
energy for the United States (2014) (Mtoe = millions of tonnes of oil
equivalent; Twh = terawatt hours; QBtu = quadrillion Btu).

The distribution of energy sources in the United States, expectedly, is
somewhat different. But still 81.2 percent is provided by fossil fuels (oil,
35.4 percent; natural gas, 28.1 percent; and coal, which has been declining, 18.3 percent), 8.3 percent by nuclear energy, and 10 percent by
renewables, the same as the global value (figure 9.3b).

If we look just at electricity usage, which begins to bring us back to
the shoreline and coastal power plants, the rankings are nearly identical

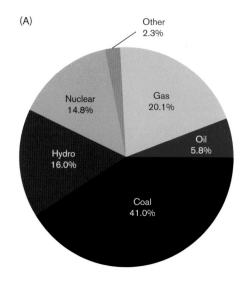

(A)

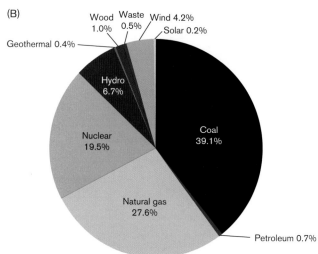

(B)

FIGURE 9.4. (a) Global sources of energy for electrical power generation (2013); (b) U.S. energy sources for producing electricity.

to total energy use, with China again no. 1, using 26 percent more electricity than no. 2, the United States. The rest of the top ten electricity-using nations are the same as those for total energy, with a little shuffling in their order. In order to produce all that electrical power, the nations of the world burn fossil fuels to provide 66.9 percent (2013) and depend on nuclear plants for 11.9 percent and renewable sources for 21.2 percent (figure 9.4a). Reaching the point where we are providing over 20 percent of global electricity from renewable sources is a very significant accomplishment. In the United States fossil fuels provide about the same percentage of our electricity (67.4 percent) as the global picture, but we are more dependent on nuclear power (19.5 percent) and less dependent so far on renewables (13.1 percent) than the rest of the world (figure 9.4b).

Our dependence on the Earth's finite supply of fossil fuels for over 80 percent of our energy needs and for two-thirds of our electrical generation cannot be sustained, particularly as usage per person in the developing regions continues to rise (why wouldn't these people aspire to the same standard of living many "developed" nations enjoy?). Some global-scale calculations have recently been made indicating that it would take about 1.6 planet Earths to support the 7.4 billion people alive today at an "average" global standard of living. It's worth thinking about that for a few minutes. What does this really mean? While we can fairly reliably get to the moon, and now even to Mars, we need to be honest. We're not going to be shipping water, minerals, and fossil fuels back from space to our home planet to support our present population, which is increasing at about 1.1 percent annually. This amounts to 75 million new people every year, or adding the entire population of both North and South Korea, somewhere, every twelve months. Looking at it another way, it's over 200,000 new human beings every day, and they all need food, water, and energy.

A large percentage of the world's people are dependent on coastal power plants, which means importing or delivering fossil or nuclear fuels to these sites, by ship, pipeline, or rail, on a regular basis. It also means contending with the impacts of the operation and by-products of the hundreds of plants that exist.

COASTAL ELECTRICAL GENERATING PLANTS

Power plants have an impact on the coastal zones in several different ways. They are a fixture along many coastlines of the United States and the world for two simple reasons: there are billions of people living in the coastal zone who need or want electricity; and the large volumes of ocean water provide the cooling necessary for the plants' operation. Inland thermoelectric power plants (those that use some source of heat to boil water, whether natural gas, coal, or uranium) must use either freshwater sources for cooling (lakes or rivers) or very large cooling

FIGURE 9.5. Diablo Canyon, on the Central California coast, is the only remaining operating nuclear power plant in California but is scheduled to close in 2025. (Photo: Kenneth and Gabrielle Adelman © 2002, California Coastal Records Project, www.californiacoastline.org)

towers, which work essentially like giant radiators. But along shorelines there is seemingly an infinite supply of cool water, salty but cool.

Although our sources of electricity and distribution lines to load centers are interconnected, where possible it makes most sense for reasons of efficiency to generate the power close to cities or other large users so there aren't significant power losses in transmitting electricity long distances. As a result, there is usually a close relationship between the locations and sizes of power plants and population densities or numbers. On the other hand, for topographic reasons, we can almost never locate hydroelectric dams near load centers, and we generally do not locate nuclear plants close to large cities.

It is useful to look at the two of the most populous states in the United States with regard to coastal populations and power plants, California (No. 1 with 39.1 million people) and Florida (No. 3 with 20.3 million people), to see what their coastal power plant picture looks like. California has 19 power plants on or adjacent to the coast, almost all of these in Southern California where the majority of the state's people live and work (figure 9.6). The greater Los Angeles area alone has about 19 million people. The state has 18 gas-fired coastal power plants and only a single operating nuclear plant (figure 9.6), with three other

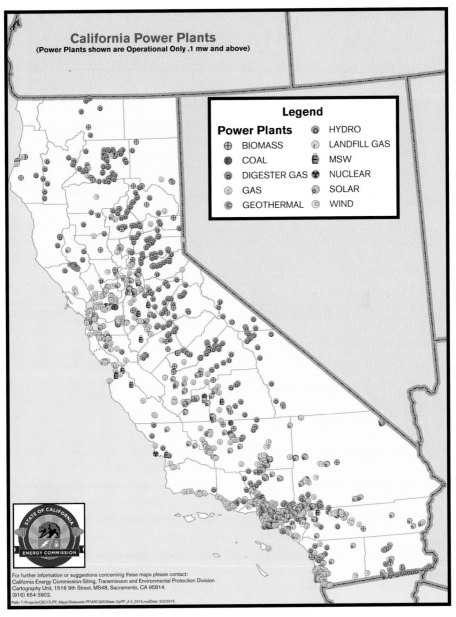

FIGURE 9.6. California has a large number of power plants, and many of the largest ones are thermoelectric plants built along the coast. (Courtesy of California Energy Commission)

nuclear stations having been closed over the years and a number of others that were proposed but never built because of public opposition and seismic concerns due to locations near active faults.

Florida has 26 coastal power plants, 14 are gas fired, 6 are coal fired, 2 use oil, 2 are nuclear, one is biomass, and one uses a mix of fossil fuels. Most of these are along the eastern coast of the state where the majority of the state's population is concentrated. All of this power plant information is readily available thanks to a very useful interactive on-line map from the U.S. Energy Information Administration, which includes every energy facility of any kind in the entire nation (http://www.eia.gov/state/maps.cfm).

It is extremely difficult, however, to get any accurate global numbers, whether people or coastal power plants. Today there are 196 individual nations scattered on six continents, and most have challenges keeping track of their own people and what they're doing. So coming up with a reliable number for something centrally important to coasts around the world—how many people live in coastal regions—for example, is no trivial task. But the most commonly cited numbers indicate that about 40 percent of the world's people (about 3 billion) live within 60 to 100 miles of a coast. Eight of the ten largest metropolitan areas on the planet, which are home to just over 200 million people, are sited along coasts. And all these hundreds of millions of people use electricity, provided, for the most part, by coastal power plants.

IMPACTS OF COASTAL POWER PLANTS

Virtually all coastal power plants are thermoelectric. They use a source of heat to boil water, and the steam generated is piped off under pressure to turn a turbine, which is connected to a generator that produces the electricity (figure 9.7). The steam is condensed though a heat exchanger, which brings cold water next to the steam. The steam is now hot water and is put back through the closed loop and back to the boiler. Through this process the cooling water from the ocean is now about 18° to 27°F (10°–15°C) warmer.

The steam-generating process is pretty simple and fundamentally goes back to James Watt, who developed the first refined steam engine in 1765, which ultimately brought on the industrial revolution. Historically, the main source of heat for producing steam was coal, which continued to be the major fuel until the early twentieth century, when

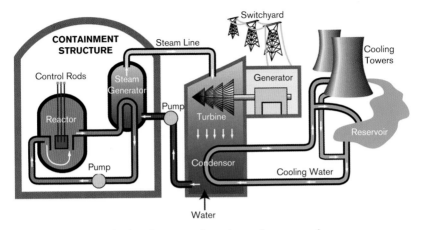

FIGURE 9.7. Flow path of cooling water through a nuclear power plant.

oil and natural gas became important fuels. Another nearly fifty years passed until nuclear energy developed as a peaceful use of the atom, but it has traveled a rockier path. After the Fukushima Daiichi nuclear plant disaster, its future is even less certain.

In any fossil fuel–fired power plant, only about one-third of the energy from burning the fuel is actually converted into electricity. The remaining energy is released as heat to the environment, whether into the coastal ocean, an estuary, a lake, a river, or another water body. (With cooling towers, the heat goes into the atmosphere.) There can be significant impacts on any body of water from this additional heat, which is referred to as *thermal pollution*.

The combination of continually increasing populations in coastal zones globally and the growing demands for electricity has intensified the need for cooling water in power generation as well as in other industrial processes. These demands have been greatest in the industrialized nations of Europe, North America, and Asia. Much of this impact has historically been felt in rivers, and in the United Kingdom, for example, it is estimated that half of all river flow is used for cooling purposes. As early as the 1980s in the United States, thermal discharges amounted to one-sixth of the total national river flow.

Coastal ocean waters are quite different, however, in large part because of the huge volume of the ocean (97 percent of all the water on the planet) and the ability for that heat to be diluted and dispersed through the action of waves, tides, and nearshore currents. Nonetheless,

FIGURE 9.8. One of the largest gas-fired thermoelectric power plants in the United States is sited at Moss Landing on the Central California coast. (Photo: Kenneth and Gabrielle Adelman © 2013, California Coastal Records Project, www. californiacoastline.org)

depending on the volume of water being released, the temperature differential between the effluent and the adjacent ocean, the depth and distance offshore for the discharge, and the circulation and mixing at that location, there can still be significant impacts.

The two largest power plants along California's coast provide some perspective on discharge volumes. The Moss Landing Power Plant (2,500 megawatt [MW] capacity, or enough electricity for about 2.5 million people), located on Monterey Bay along the Central Coast, is one of the largest natural gas–fired plants in the nation (figure 9.8). The plant takes in about 1.2 billion gallons of water a day from the adjacent Elkhorn Slough and discharges it 650 feet offshore in Monterey Bay about 20°F (11°C) warmer than the original intake water. The Diablo Canyon Plant (2,240 MW, or enough power for about 2.3 million people) is a nuclear generating station, the only one still operating in California, and it discharges about 2.2 billion gallons per day of cooling water into the adjacent ocean, which is about 20°F warmer than the intake water. For the state as a whole, power plants suck in about 15 billion gallons of seawater for cooling every day.

Heating the water through the cooling process, whether along the coast of Florida, California, or along any water body, can produce a number of direct and indirect effects, including the following:

1. *Reduction of dissolved oxygen content:* Warmer water holds less dissolved gas than cold water, so as water temperature is increased, dissolved gases are given off. Higher organisms, such as fish, need a certain amount of dissolved oxygen, so this reduction can be problematic at certain levels, depending on the type of organism.

2. *Direct thermal effects:* Because all marine organisms, except marine mammals, are cold-blooded, when the surrounding water temperature changes, the body temperatures of the organisms also change. In many cases, the faunal population around the vicinity of an outfall or discharge point may simply change as certain fish and other organisms migrate away from the warmer water and new species are attracted. Depending on the initial ambient water temperature, which can be in the 80° or even 90° range in a place like Florida, an increase in 20°F can result in thermal stress that is lethal to some species. With fish, for example, there are many variables such as previous temperature exposure, age, diet, season, and chemical composition of the water that may alter the specific lethal temperature point. Eggs, larvae, or juveniles of any organism are particularly vulnerable to small changes in water temperature. While 20° F may not initially sound like a huge difference if you are considering humans and our surrounding air temperature (say, a change from 70° to 90°), it is quite different when your surrounding medium is water. A better comparison is to imagine sitting in a hot tub or hot spring of a moderately comfortable 102°F and then increasing it to 122°.

3. *Increase in toxicity:* Synergism describes the combined effects of two or more environmental changes or contaminants. The physiological effects of certain toxins in the receiving water can be increased with higher water temperatures.

4. *Altering metabolic and other physiological processes:* Water temperature exerts an important influence on many basic biological processes, including metabolism, and biochemical processes that control respiratory rates and digestion. Substantial ocean water temperature changes have the potential to cause significant organism or ecosystem disruption.

5. *Reproductive impacts:* Even relatively small increases in temperature can have major effects on reproduction in marine organisms. Water temperature is usually seasonally controlled, and temperature thresholds exist for release of eggs for shellfish, for example, and can therefore affect hatching time and success, as well as spawning and migration.

Two additional impacts of the initial intake of large volumes of cooling water from the coastal ocean that have been known for a long time are the *impingement* and *entrainment* of marine organisms in the water intake system. With up to two billion gallons of seawater per day being sucked into a large pipe or pipes, the potential impact can quickly be appreciated. The most common solution for reducing both of these potential impacts, at least to intermediate-size or larger animals, is to use a screened intake, with the mesh size of the screen determining what size marine life will either be caught on the screen (impingement) or pass through the cooling water system (entrainment). The screens utilized on power plant intakes have a small enough screen size that marine mammals and fish are not pulled into the intake.

In July 2015 two friends were scuba diving while boating near the St. Lucie nuclear power plant in South Florida. They discovered three massive barnacle-covered structures beneath the surface, which looked like some interesting underwater building. There were no warning signs, but as they swam up next to one of these pipes, one of the men felt a current, and before he knew what happened, he was sucked into a large pipe along with thousands of gallons of seawater. For a quarter of a mile he was pulled along in a pipe 16 feet in diameter in a current that became increasingly turbulent. Just as he imagined the worst, he was dumped out into a large intake pool and was discovered and rescued by a plant worker. It is of some concern that the exact same thing happened to another scuba diver in 1989 at the same location. A very similar incident occurred along the coast of Playa del Rey, in Southern California, when a scuba diver fishing for lobster was swept into the mouth of a huge intake pipe for the Los Angeles Department of Water and Power's Scattergood power plant. He also fortunately ended up in a catch basin and was none the worse for his trip.

This is an infrequent occurrence, and the simple solution is to have a screen over the intake, which is usually the case. Another approach that can be used in combination is to employ several different intakes in order to reduce the flow into each pipe, so that swimming animals can more easily avoid the volume and velocity of the inflow at any one intake. The marine life that will be entrained in an intake structure will normally be the microscopic plankton, larvae, and fish eggs, which do not survive the transit through the cooling water cycle. There are literally trillions of these small organisms in the surface waters of the oceans, and studies have been carried out to assess the magnitude of the loss, which normally tends to be very localized.

One approach to reduce these planktonic entrainment losses is to extend the intake pipe farther offshore and pump in water below the photic zone, which is generally defined as the upper 100 to 300 feet of the ocean and which contains most of the microscopic marine life. In places, however, Monterey Bay, California, for example, the photic zone may only extend to a depth of about 100 to 150 feet. A deeper intake would, however, add very significant costs, as these depths would normally be several miles offshore, unless there is a deep submarine canyon nearby, which would provide deep water very close to shore.

Federal rules have now banned new plants from drawing in seawater through what are known as *once-through cooling systems,* and California is proposing a similar set of requirements for the state's coastal power plants. Another proposed approach is to eliminate approval of once-through cooling for new power plants but to allow twelve years for existing plants to comply. Eliminating once-through cooling would probably require the construction of very large cooling towers, which have their own environmental impacts. Cooling ponds can also be used; they would not require the construction of cooling towers but would require adequate land area for pond construction.

It is important to reach some clear consensus based on good science on all of the environmental impacts of different cooling water systems before the planning and construction of new power plants or altering the practices employed by existing plants. While the loss of large numbers of plankton may initially seem like an unacceptable outcome, it is important to put this in perspective. What have been the impacts on coastal fish populations, for example, of the existing power plant cooling water systems relative to other causes of mortality? How much area is affected by plankton losses? Field studies conducted around the outfall of the very large Moss Landing Power Plant (figure 9.8) on the Central California coast (1.2 billion gallons/day intake/discharge volume) indicated that there was an estimated 13 to 28 percent loss of larvae in an area of 390 to 480 acres of the adjacent ocean. This is not intended to completely disregard the importance of plankton but to put in perspective the loss of a very small percentage of a massive amount of microscopic life in the coastal ocean, only a very, very small percentage of which lives to maturity. A female codfish, for example, lays between 4 million and 6 million eggs at a single spawning. All but a handful of these will end up as food for someone else out there.

What are all of the environmental impacts of constructing massive concrete cooling towers? Making cement involves the release of large amounts of carbon dioxide to the atmosphere, with one ton of cement adding one more ton of $CO_2$ to the atmosphere. After the combustion of coal, oil, and natural gas, the production of cement is next in global $CO_2$ generation. We are faced with many difficult environmental decisions today because everything is connected to everything else. While requiring some new level of treatment or some new approach, such as the elimination of once-through cooling water for coastal power plants, may initially seem like the right choice or decision, it may in fact have greater overall negative environmental impacts than the original process we are trying to mitigate.

One potential benefit of releasing warmer water at depth offshore is that this can induce artificial upwelling of nutrient-rich bottom waters, thereby stimulating productivity in the surface waters. This has been proposed as a way to encourage aquaculture or fish farming in some coastal waters.

## NUCLEAR POWER PLANTS

The main difference between a thermoelectric plant powered by fossil fuel and one powered by uranium is that with the latter, the fuel and the potential impacts of its extraction from the Earth, refinement, transport, use, release, and long-term storage need to be very, very carefully considered, managed, and controlled. That has not always been the case.

Globally, nuclear power generated nearly 12 percent of total electricity in 2013, although many of the plants are not located on the coast. France has been the world leader, with 75 percent of its electrical power generated by nuclear energy, followed by Lithuania (72 percent), Belgium (54 percent), Slovakia (52 percent), Hungary (50 percent), Ukraine (49 percent), Sweden (40 percent), Switzerland (40 percent), and trailing off from there.

In the United States 19.5 percent of our electricity is generated by 99 nuclear reactors at 65 different plants, with all but 3 of these in the East or Midwest. Despite this relatively low percentage of electricity from nuclear energy, we have the largest number of nuclear reactors of any nation on Earth, and in 2013 they generated 44 percent of the world's nuclear electricity. As of 2015 there were 5 new reactors under construction in the United States, while 33 reactors have been permanently shut down. In California alone, because of rapid growth from the 1950s on, 6 large nuclear plants were proposed, but they were never built because of public opposition, due in large part to questions regarding locations near active faults and therefore seismic safety. Three others were built, operated, and then closed, leaving only one operating plant, Diablo Canyon, on a remote section of the San Luis Obispo County coast of Central California (see figure 9.6), which has had its own ongoing seismic concerns. In 2016 agreement was reached to close the thirty-one-year-old plant in 2025.

The history of nuclear power is a checkered one, with early promises of "infinite energy" and a source of electricity that was "efficient, economic, clean and safe," none of which was actually true. Early development of nuclear power plants in the United States took place under the control of the Atomic Energy Commission (AEC), an agency that was placed in the highly questionable role of both promoting and regulating the use of nuclear energy after World War II. In 1975, following the passage of the Energy Reorganization Act of 1974, the Nuclear Regulatory Commission (NRC) was established to protect public health and safety related to nuclear energy. Its role included oversight of reactor safety and security, reactor licensing and renewal, and management of radioactive materials, including spent reactor fuel, its storage, recycling, and disposal. The intent of the legislation and the establishment of the NRC were laudatory. A dozen years after its creation, however, a congressional report concluded that the agency "has not maintained an arms length regulatory posture with the commercial nuclear power industry . . . [and] has, in some critical areas, abdicated its role as a regulator altogether." This was clearly an agency that wasn't doing the job it was created to perform.

Unlike other forms of electrical power generation, particularly along coastlines, where plants are intimately connected to the global ocean through their cooling water systems, there is absolutely no room for error. And human errors as well as natural disasters have occurred with tragic short-term consequences, as well as chronic long-term impacts. Humans will always make mistakes, and once radiation is released into the water or atmosphere, there is no way to get it back. There have now been three very large nuclear power plant accidents, with well-known names, Three Mile Island, Chernobyl, and Fukushima Daiichi. Three Mile Island, near Middletown, Pennsylvania, suffered a partial meltdown on March 28,

1979, and was the most serious accidents in U.S. commercial nuclear power plant history. The failure was a result of design-related problems, equipment malfunctions, and worker errors and led to what the NRC described as "very small off-site releases of radioactivity." The incident led to major changes in reactor operating training, radiation protection, emergency response training, and tightening of NRC regulatory oversight.

The combination of a flawed reactor design and serious mistakes made by inadequately trained personnel led to the disastrous Chernobyl accident in Ukraine in April 1986. One reactor was destroyed by the explosion, which led to ten days of radiation release into the environment, the largest uncontrolled radiation release of any civilian operation in history. The radiation led to the deaths of thirty workmen and firemen within the first three months and acute radiation poisoning of many others, as well as radioactivity fallout and contamination over much of northern Europe.

On March 11, 2011, a Friday afternoon, at 2:46 P.M., the first of three sequential disasters struck the northeast coast of Japan. The first was a massive 9.0 magnitude earthquake generated when the Pacific Plate slid west beneath the Eurasian Plate along a subduction zone 45 miles offshore. This was the largest earthquake to hit Japan in recent history and the fourth most powerful earthquake in the world since modern record keeping began in 1900.

The seafloor displacement of 20 to 25 feet created a massive tsunami, which hit the coastline of Japan's northern islands in less than an hour. This was disaster No. 2. The surge reached elevations of up to 128 feet above sea level and moved inland as far as 6 miles, flooding over 200 square miles of low-lying coastal land (see figures 2.8 and 2.9). The earthquake and tsunami led to the deaths of nearly 21,000 people, mostly from drowning. This was neither the largest nor the deadliest earthquake and tsunami to strike this century, however. That unfortunate record goes to the 2004 magnitude 9.1 Sumatra events, which killed more than 230,000 people living around the Indian Ocean.

Although four Japanese nuclear power plants were automatically shut down following the 2011 earthquake, the reactors still required cooling water to remove heat. At the Fukushima Daiichi nuclear plant, tsunami waves overtopped a 33-foot-high seawall protecting the diesel backup cooling facility, flooding and disabling the system. The loss of cooling water led to three large explosions followed by nuclear meltdowns at three of the plant's six reactors. Radioactivity was released from the reactor containment vessels due to uncontrolled leakage but also from deliberate

FIGURE 9.9. Millions of gallons of contaminated water are stored in tanks surrounding the Fukushima Daiichi nuclear plant. (Imagery © 2016 Google, TerraMetrics, map data © 2016 ZENRIN)

venting to the atmosphere to reduce pressure and from deliberate discharge of coolant water to the adjacent ocean. This started disaster No. 3.

Combined with the approximately 440 tons per day of cooling water that was pumped in to cool the reactors, an additional estimated 300 tons of groundwater has been flowing beneath the reactors daily, picking up radiation and carrying it to the adjacent ocean. Over the years since the accident, over a thousand large tanks were set up in an attempt to collect the contaminated water (figure 9.9). Treatment facilities were also constructed in an effort to partially clean the water.

The complications are ongoing. One troubling problem has been how to cut off the subsurface flow of groundwater beneath the plant, which mixes with radioactive water leaking from the reactors and flows into the ocean. The plan that was finally developed, nearly three years after the earthquake, tsunami, and meltdown, was to freeze the ground and groundwater beneath the site and build an ice dam to contain the flow. This complex project was supposed to be completed in March 2015, four years after the initial disaster, but offshore radiation measurements in late 2015 indicated that while the flow had been reduced, it had not yet been completely eliminated.

This disaster has taken a large toll on the outlook for greater dependence on nuclear power. There were no new nuclear electrical generating plants built in the United States for about thirty years following the Three Mile Island accident in 1979. In 2012 four new reactors were approved for construction by the NRC as a result of increasing awareness of the impacts of burning fossil fuels on global warming and climate change. The explosions and meltdowns of the reactors at Fukushima Daiichi and the release of large amounts of radiation, both into the atmosphere and into the ocean, have had a major impact on public perceptions and also government policies globally. Prior to Fukushima, Japan relied on nuclear energy for about 25 percent of its electrical production, but most of the reactors were subsequently shut down because of safety concerns. In January 2013 most cities hosting nuclear plants stated that they do not mind restarting nuclear power plants if the government could guarantee their safety (which is essentially impossible to do). Three years later, in February 2016, over 70 percent of the Japanese people were in favor of completely or partially abandoning the use of their nation's nuclear power plants, and only 3 percent felt that the nation should continue to build new plants.

In addition to Japan's immediate radiation releases, concerns along the U.S. West Coast about potential radiation exposure arose soon after the release of radiation from the plant was reported. Traces of radiation in the atmosphere were detected over the area within five days as atmospheric circulation carried the releases around the world.

The greatest amount of radiation (primarily cesium and iodine) entered the ocean off Japan in the first several months after the reactor failures. It was described as the greatest individual emission of artificial radiation into the sea ever observed. In a survey of 170 different types of fish caught off Japan, 42 species tested in the months immediately after the accident had too much radiation for consumption. Ken Buesseler, a radiochemist from the Woods Hole Oceanographic Institution, has been systematically sampling the waters off Japan and across the North Pacific since the accident. By October 2015, four and a half years after the meltdown, the levels of cesium off Japan are "thousands of times lower" than during the weeks immediately following the incident, yet still not "under control." To put the values measured in perspective, however, Buesseler reported that the radiation levels detected in the coastal water near the reactors in late 2015 were still more than forty times lower than U.S.

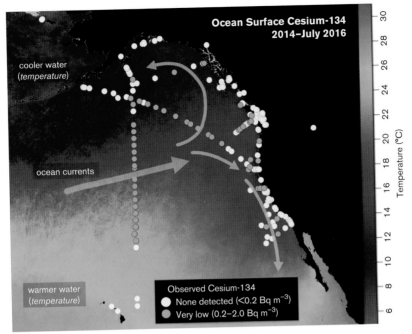

FIGURE 9.10. Locations of seawater samples analyzed at the Woods Hole Oceanographic Institution for radioactive cesium. Orange dots indicate cesium-134 contamination from the Fukushima Daiichi nuclear plant. (Courtesy of Ken Buesseler, Woods Hole Oceanographic Institution)

safety standards for drinking water, and well below limits of concern for direct exposure from swimming or other ocean recreation.

Without question, the greatest concern with any coastal nuclear plant accident is the release of radiation to the immediate area. Ocean circulation will carry the radioactive elements globally, however, although at gradually diluted levels and with progressively less radioactivity over time because of the decay of the short-lived radioactive elements like cesium. Circulation across the North Pacific from Japan to the West Coast is a relatively slow process. This is a 5,000- to 6,000-mile trip, and with typical Kuroshio Current and North Pacific Drift current speeds, model calculations indicated the radiation would begin to arrive off the west coast of North America in 2014 and peak in 2016. Buesseler and his research team first documented cesium along the West Coast in February 2015 off British Columbia, four years after the initial radiation release (figure 9.10). The highest values were reported 1,500 miles north of Hawai'i. If you were to swim in this water, the dosage or health effects (based on the level of radioactivity measured in Becquerels/cubic meter of seawater, with one Becquerel equal to one decay event/second) have been described as a thousand times less than a single dental X-ray.

Additional samples off the U.S. West Coast have detected cesium from the Fukushima Daiichi plant but at levels more than five hundred times lower than U.S. government drinking water standards. These values are consistent with those measured by Canadian scientists, who reported they have found no cesium-134 (from the plant) in fish collected off British Columbia. The transit across the North Pacific has been quite slow, averaging just several miles a day. This has given the radioactive cesium from the damaged plant, which has a half-life of just two years, additional time to decay and dissipate.

Buesseler said these data are important for two reasons: "First, despite the fact that the levels of contamination off our shores remain well below government-established safety limits for human health or to marine life, the changing values underscore the need to more closely monitor contamination levels across the Pacific. Second, these long-lived radioisotopes will serve as markers for years to come for scientists studying ocean currents and mixing in coastal and offshore waters."

### WHERE DO WE GO FROM HERE?

We continue to demand and use large amounts of electrical power, much of which is generated by thermoelectric power plants located along coastlines where the load centers are located and where essentially unlimited amounts of ocean water are available for cooling. However, there are impacts associated with these plants, whether fossil fuel fired or nuclear. For nuclear plants, the concern is the long-term effects of radiation from an accident at any stage in the uranium cycle. The burning of coal and natural gas and the emission of carbon dioxide into the atmosphere, and also uptake by the ocean, produce the greatest long-term impacts, which are being increasingly felt globally. While we can reduce the impacts of thermal discharges from either type of coastal power plant and continue to try to improve nuclear plant safety, our ultimate goal should be to make the transition to renewable energy sources as rapidly as possible through all types of incentives and encouragement that government agencies at every level around the planet can provide. It is not going to be easy, but, quite frankly, for our own survival, we have no other choice.

CHAPTER 10

# Renewable Energy from the Coastal Zone

Our planet's 7.4 billion people use a lot of energy, and some of us use a lot more than others. Whether we measure it in barrels of oil equivalent or terawatt hours per year (TWh/yr), it's a lot of energy, and the demand is increasing. The global use of energy increased 350 percent in the fifty years between 1965 and 2014, or on average about a 20 percent increase every decade. Until very recently, most of the increase was in fossil fuels—coal, oil, and natural gas—which still provide about 85 percent of our global energy. Renewables make up an increasing proportion, now about 9.4 percent of global energy use, and are used to generate 18.3 percent of the electricity (see figures 9.3, 9.4). In the United States renewable sources provide 9.8 percent of the total energy and 13.1 percent of the electricity.

We know that coal, oil, and gas took millions of years to form, and what we have now is all we are ever going to have. Fossil fuels are nonrenewable resources whose use is nonsustainable. The same can be said for nuclear energy, although it may have a longer time frame. As energy prices have gradually increased (when I started driving in 1958, gasoline was about 25 cents a gallon) and as the readily accessible oil, natural gas, and coal resources have been exhausted, the petroleum industry has gone deeper into the Earth, offshore into deeper and deeper water, and also ventured into the Arctic. And we are now extracting petroleum from oil shale and tar sands and also hydrofracturing the bedrock to allow for the recovery of more oil and gas. Not only do all these processes involve a number of significant environmental impacts and risks, but burning fossil fuels has had well-documented effects on climate change.

We are not yet running completely out of fossil fuels, but we are running out of atmosphere. The Stone Age didn't end because we ran out of stones but because humans discovered better ways of doing things. The same can be said for fossil fuels. We know there are limits to what is still in the ground, although there is not complete agreement on just how much is still buried in the Earth, and also that there are very significant impacts from burning more fossil fuels. Sooner or later, we are going to have to rely on renewable or sustainable sources of energy, because the oil, gas, and coal are going to be exhausted. Not today, not next year, but it's going to happen, and the sooner we make a major shift to renewables, the more of the planet and its atmosphere, ocean, and climate we can stabilize and preserve.

COASTAL RENEWABLE ENERGY SOURCES

The conventional list of existing sources of renewable energy includes hydropower, geothermal, biomass or biofuels, wind, solar, and hydrokinetic. Moving to the coastal zone, however, whether onshore or offshore, the list gets reduced, with wind being the only one concentrated along coastlines. But we can also add *hydrokinetic* energy (power from tides, waves, or ocean currents) and ocean thermal energy conversion.

*Use of Wind Power*

Wind is really just the movement of air from an area of high pressure to an area of low pressure. Because solar energy heats different areas of the earth unevenly, warm air will rise where the land or ocean has been heated, leaving a void, and the movement of an air mass into that void creates wind. Certain regions of the Earth have very regular wind patterns as a result of reasonably consistent temperature differences, and coastal areas provide good illustrations of this process. California's interior valleys, for example, typically warm up during the sunny morning and midday hours of summer and fall, causing the warm air to rise, which then generates afternoon winds that tend to blow from the coast into the inland valleys.

Wind energy has been used for centuries, initially by the first mariners who hoisted a sail on a mast to power a ship. Windmills have been used to pump water from the ground across the American West and Australian Outback, where there was no other convenient or available source of power. The first windmills used to generate electricity were constructed in 1887 in Scotland and in the United States, in Ohio. These were used for lighting a few buildings, but it wasn't until the early years of the twentieth century and the development of electrical generating plants and transmission lines that wind power had a way to transmit its benefits. It has grown the most rapidly over the past twenty-five years of all renewable energy sources, and today the wind energy industry is booming. Globally, generation more than quadrupled between 2000 and 2006, then quadrupled again between 2006 and 2012, although it still provides a relatively small amount of the world's total electrical energy, 4.7 percent of U.S. electricity and 4 percent of global electrical power.

The United States was an early leader in the development of commercial wind energy and had the largest installed capacity until the 1990s. Germany began investing heavily in wind energy, however, and took over the lead in about 1997, which lasted a decade. The United States expanded its capacity and by 2008 was again the world leader. China invested heavily in wind power, and by 2010 it became the global leader. By 2011 eighty-three nations were using commercial wind power to generate electricity. By 2014 China had installed 31.1 percent of the total global wind power; the United States was second with 17.7 percent, followed by Germany (10.8 percent), Spain (6.2 percent), India (6.1 percent), the United Kingdom (3.4 percent), Canada (2.6 percent), France (2.5 percent), Italy (2.3 percent), Brazil (1.5 percent), and the rest of the world 15.8 percent.

Total installed global wind capacity by the end of 2016 was 82,183 megawatts, which was generated by over 50,000 individual wind turbines. If we use the very high U.S. electrical consumption rates (1 MW supplies power to about 250 homes, or about 1,000 people), wind energy could provide for the electrical needs of 108 million households or about 432 million people, 5.8 percent of the Earth's population. However, most of the world's people use far less electrical power than do Americans, so the actual number of people served is no doubt significantly greater, and global capacity has been doubling every three years.

Most of the wind power being generated today comes from wind turbines, which look like huge airplane propellers (figure 10.1). The individual blades can be 200 feet in length, and a typical turbine is about the height of a twenty-story building. The rotation of the blades from the movement of the wind turns a shaft that is connected to a generator that

FIGURE 10.1. A large wind farm in Spain, one of many such installations across Europe. (Photo: D. Shrestha Ross © 2015)

produces the electricity. A wind turbine is exactly the opposite of a household fan. The fan uses electricity to make wind, and the wind turbine uses the wind to make electricity. A single large turbine can produce enough electricity for the needs of about 2,400 Americans, or about 600 households. This same turbine could power the homes of many more people in almost any other nation on the planet, simply because per capita uses are considerably less. Typically large numbers of these turbines, known as wind farms, are erected in places where wind is relatively consistent and strong and construction permits can be approved.

In contrast to fossil fuels and nuclear power, which at present provide about 87 percent of all electricity in the United States and 82 percent globally, wind is a clean source of renewable energy that produces no air or water pollution and no greenhouse gases, and the land where wind farms are constructed can also be used for other purposes (agriculture and grazing, for example). Because wind is free, operational costs are low relative to many other forms of energy once a turbine is purchased and erected. Advancements continue to be made in the technology and manufacturing of turbines, so costs are coming down and many governments have offered tax incentives to encourage wind energy development.

The single most important reason for our continuing dependence on fossil fuels for the great majority of our electrical generation is simple economics, which to a large extent has been influenced by politics. Coal and natural gas still provide what, at least at the surface, is the least expensive source of electricity. While the cost of wind power has come down substantially in the past ten years, it is still a higher-cost alternative because the technology requires a greater initial investment per MW than fossil fuel–fired power plants. For today's wind turbines, about 80 percent of the startup cost is in the machinery, with the other 20 percent involved in site preparation and installation.

However, simply looking at the initial investment has been shown not to be a good measure of actual costs. This is a much more complicated analysis if we look at all of the other costs. If wind turbines are compared to coal, gas, or nuclear power plants on a "life-cycle" cost basis (which has been quantified as the projected levelized cost of electricity, or LCOE)—which includes the average total cost to build and operate a system (including fuel costs) for the life of the facility divided by the total energy produced—wind energy is much more competitive with these other technologies because there are no fuel costs (which can fluctuate widely) and only minimal operating expenses. The projected LCOE in the United States for 2020, using average $/MWh, puts advanced natural gas systems ($73/MWh) and onshore wind energy ($74) as the least expensive, followed by hydroelectric power ($84), conventional coal and advanced nuclear (both at $95), biomass ($101), coal gasification ($116), solar photovoltaic ($125), conventional natural gas ($142), offshore wind ($197), and solar thermal ($240) (figure 10.2).

The other very large costs, which are not factored into even this accounting metric, are what are labeled external costs, which are all of the costs that are indirectly passed on to society as a consequence of producing electricity from a particular energy source. The biggest of these, and it dwarfs all others, is the combined environmental impacts of increased atmospheric greenhouse gases from the combustion of coal and natural gas: increased rates of sea-level rise and impacts on shorelines and nations around the world, ocean acidification (see chapter 13), human health issues (not only global, but local with coal mining), climate change, and effects on water supplies and food production. A price or tax on carbon emissions has been proposed, and is being implemented in some areas, as a way to begin to recover some of these global societal costs. The Fukushima Daiichi disaster comes to mind as a very large external cost for producing electrical energy from a nuclear plant, with total damage and cleanup costs now estimated by the government of Japan at about $250 billion. This one accident changes the economic picture of nuclear power in a very significant way.

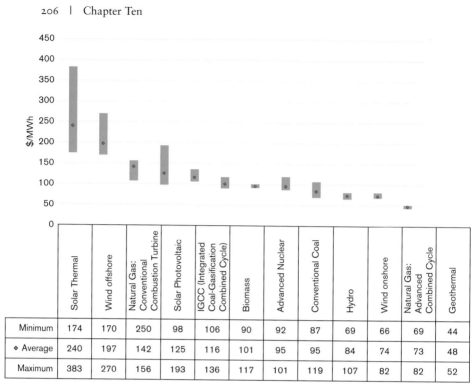

| | Solar Thermal | Wind offshore | Natural Gas: Conventional Combustion Turbine | Solar Photovoltaic | IGCC (Integrated Coal-Gasification Combined Cycle) | Biomass | Advanced Nuclear | Conventional Coal | Hydro | Wind onshore | Natural Gas: Advanced Combined Cycle | Geothermal |
|---|---|---|---|---|---|---|---|---|---|---|---|---|
| Minimum | 174 | 170 | 250 | 98 | 106 | 90 | 92 | 87 | 69 | 66 | 69 | 44 |
| ◆ Average | 240 | 197 | 142 | 125 | 116 | 101 | 95 | 95 | 84 | 74 | 73 | 48 |
| Maximum | 383 | 270 | 156 | 193 | 136 | 117 | 101 | 119 | 107 | 82 | 82 | 52 |

FIGURE 10.2. The projected levelized cost of electricity (LCOE) production is a better measure of comparing all of the costs associated with different methods of electrical generation. $/MWh = $/Megawatt-hour. (United States Energy Administration)

## Offshore Wind Energy

While a number of wind farms have been sited along coastlines, placing the wind turbines offshore has the advantages of not taking up valuable coastal land and access to stronger winds and, therefore, the potential to generate larger amounts of electricity. An additional benefit is that there is usually but not always, as the United States has learned, less public opposition to offshore wind farms. In addition, due to the large coastal concentrations of people around the world, these wind farms can be sited close to electrical grids and, therefore, eliminate overland transmission line costs and power losses. On the other hand, placement of offshore wind turbines involves significantly greater construction costs and has projected levelized costs of electricity about two and a half times greater than wind turbines placed on adjacent land. The turbine itself makes up about one-third to one-half the cost of offshore installations, with infrastructure, maintenance, and oversight making up the rest. To place an offshore wind turbine

FIGURE 10.3. The Block Island wind farm project off the coast of Rhode Island was completed in 2016. It is the first wind farm to be constructed in U.S. waters. (Photo: Courtesy of Deepwater Wind)

with 200-foot-long blades requires some extraordinary machinery, including a very large ship, which is specially designed to transport and erect each of the turbines and its foundation (figure 10.3).

Denmark installed the first offshore wind farm in 1991, and by the end of 2014 eleven European countries had installed 3,230 offshore turbines at eighty-four individual wind farms having a total capacity of 11,000 MW. Five more large facilities are under construction offshore of the Netherlands, Germany, and the United Kingdom, and ten more even larger wind farms are in the planning and permitting stages. The United Kingdom has by far the largest offshore operating capacity with 3,681 MW, followed by Denmark (1,271 MW), Belgium (571 MW), Germany (520 MW), and the Netherlands and Sweden. The three largest operational wind farms in the world are located off the coast of the United Kingdom, and each has between 140 and 175 individual turbines (figure 10.4).

The total operating capacity of offshore wind farms more than doubled between 2006 and 2009, then quadrupled between 2009 and 2014. Northern Europe has been leading the world in both existing capacity and building and planning new wind farms. Although offshore wind turbines have been successfully providing power to northern Europe since 1991, there is only a single operating offshore wind farm along the

FIGURE 10.4. Offshore wind farm off the southeastern coast of the United Kingdom. (Photo: D. Shrestha Ross © 2015)

entire U.S. coastline. With the exception of the Block Island, Rhode Island, project, all efforts to date to put a wind farm in the water have failed because of a combination of projected or perceived costs, bureaucratic challenges, political opposition, and environmental concerns, including the visual impacts of the tall towers and the potential for impacts on birds and marine mammals. There has now been ample experience with European offshore wind turbines, however, as well as considerable baseline information gathered along the U.S. mid-Atlantic offshore area on birds, sea turtles, and marine mammals, so that many of these problems have been resolved or greatly reduced through technology refinements and optimal siting of the farms. We need to build on this existing foundation of research, environmental impact assessment, and experience with existing facilities and not see every new proposal as the time to start all over again. While any large energy facility will be visible from somewhere and there will always be some local opposition, we need to look carefully at the alternatives and their impacts. As long as we continue to oppose and delay well-planned renewable energy projects, the more fossil fuels we will continue to burn and the more greenhouse gas we will emit, with all of their impacts.

The Cape Wind Project.

The proposed Cape Wind, Massachusetts, project is a good example of the complexity of the challenges and difficulties that lie along the path to successful approval and construction of an offshore wind farm in state waters in the United States. For ten years, beginning in November 2001, when the original application was submitted, this project was expected to be the first U.S. offshore installation. Relatively calm and shallow water and the consistent winds seemed to make the area off Nantucket Sound an ideal location. The proposal involved the construction of 130 individual turbines about 5 miles offshore, each about 250 feet high, covering an area of about 25 square miles. At completion the project would have had a capacity of 468 megawatts, or able to provide for about 117,000 households, roughly 468,000 people. But the project faced strong opposition from a number of organizations, which brought dozens of lawsuits against it, claiming that it would harm birds and other wildlife, increase electricity rates, endanger airplanes, drive tourists away, conflict with fishing and recreation, industrialize the area, damage submerged vegetation, change sediment transport, and fragment habitat—this was a very long list— but it appears that the greatest concern for residents of Martha's Vineyard and Nantucket was that it would interfere with their ocean views.

The wind farm did break some important new ground in being the first U.S. offshore wind farm proposal to conduct an extensive environmental assessment. The thousands of pages of that independent analysis helped to calm some groups that were skeptical of the project. Except for one temporary decision, all of the judicial rulings were in favor of the Cape Wind project. Cape Wind Associates actually received the first commercial offshore renewable energy lease in the United States, which became effective November 1, 2010. Following the submittal and approval of additional documents, but with continued opposition, Cape Wind Associates requested a two-year suspension of the operating terms of its lease in February 2015, in part due to the challenges of financing the $2.5 billion project. Its future as of 2016 is unclear.

Deepwater Wind: The Rhode Island Project.

Construction of America's first offshore wind farm, about 85 miles southwest of the proposed Cape Wind project site and 3 miles off the coast of Block Island, was completed in 2016. Deepwater Wind built the facility, which includes five turbines generating 30 megawatts of electricity (Figure 10.3). The turbines will supply electricity for the island community's needs, although it is considerably smaller than the proposed

130-turbine Cape Wind facility. The island has always relied on a costly diesel generator system, and the wind turbines will reduce the island's electricity rates by an estimated 40 percent and also diversify Rhode Island's power supply with renewable energy. Several major differences in this project are important to consider for future proposals for offshore wind farms anywhere: by starting smaller, the project was easier to finance; and the offshore location selected had already been designated for this type of development by the State of Rhode Island and had built-in government and local support.

## The Future of Offshore Wind Farms

The U.S. Department of Energy launched the National Offshore Wind Strategy in 2011 and has allocated over $285 million for offshore wind technology development, demonstration projects, and market acceleration. With 80 percent of the nation's electrical demand coming from coastal states, there is a strong belief that offshore wind energy can help meet this demand with clean and renewable energy. A 2015 report released by the Energy Department titled "Wind Vision" (http://energy .gov/sites/prod/files/WindVision_Report_final.pdf) provides a detailed analysis of the future for wind power in the United States. The potential is large, and by 2050 the offshore wind industry could employ 170,000 to 180,000 people and contribute up to 86,000 MW of power (enough for about 86 million people).

As of early 2016 there were twenty-one offshore projects in the planning and development stages, and thirteen of these were in the more advanced stages, totaling nearly 6,000 MW, enough to generate electricity to power about 1.5 million homes and provide for the needs of about 6 million people. Potentials are one thing; permits, financing, construction, and operation are quite another. The economic costs of offshore wind still remain significantly higher than onshore wind farms, for reasons discussed earlier, but again, the true societal costs of generating plants powered by fossil fuels have not to date been included in energy cost comparisons.

In addition, as the Cape Wind proposal clearly demonstrated, there are major regulatory challenges, because of a system that often does not provide pathways, incentives, or sufficient certainty for wind energy companies to embark on costly planning processes with no clear endpoint or approval in sight. Rhode Island has set one good example, and zoning or designating specific offshore areas for wind farms following thorough environmental assessments is a logical starting point. While there are clearly regional differences in marine life and living marine resources that must be considered, we now have twenty-five years of offshore wind farm experience in European waters off eleven different countries. With 3,230 offshore turbines in place at eighty-four wind farms, there ought to be enough information from these facilities to help inform any proposals in U.S. waters. We aren't starting from scratch here, and we don't need to completely reinvent the wheel with each new proposal.

Supporters of offshore wind, and the industry itself, praise the energy policies and financial incentives of the European countries that have moved aggressively forward with the development of offshore wind energy. Denmark, for example, has a goal of providing 50 percent of its power from wind by 2020, just three years away. In order to help achieve that goal, the government is setting a price for electricity from those facilities and also requiring that offshore wind turbines be connected to the power grid. This type of long-term encouragement and these policies are considerably more attractive to the wind energy industry than the short-term tax credits that the U.S. government has provided to date, which can also disappear with a new election cycle. These may be issues that are better managed at the state level, however, where ocean zoning, tax incentives, and government encouragement and support may be easier to develop.

Coastal wind energy, whether onshore or offshore, is going to be with us forever. It is never going to run out, and while there are substantial costs involved in developing a wind farm, the operational costs aren't subject to the uncertainties in fuel costs that thermoelectric plants must cope with. There are certainly daily fluctuations and seasonal variations, but offshore wind in many areas is relatively consistent and predictable so that wind energy projects have some certainty in the resource, if not a clear route to developing it. While cost issues remain, there are significant external costs borne by society as a whole to U.S. and global dependence on fossil fuels that need to change, whether carbon taxes or some other mechanism.

## Hydrokinetic Power

Offshore marine and hydrokinetic (MHK) technologies are still in their infancy, but the potentials are very large, although as discussed earlier there is sometimes a huge gap between potential and implementation. Marine hydrokinetic includes harnessing tidal energy, ocean currents, and waves. As long as humans have watched the tides roll in and out every day, particularly in bays and estuaries where the daily range between high and low tides may be 8 or 10 feet or much more, there have been those who have imagined harnessing all that energy. Powerful waves breaking along the shoreline have invoked the same responses. With more than 50 percent of the U.S. population living within 50 miles of the coast, off-shore marine and hydrokinetic as well as wind energy could provide a significant amount of electrical energy next to areas where it is most needed. And the global population is similarly concentrated in coastal regions. So what is the potential? Have we made any progress, and are there success stories of utilizing wave, tidal, or current energy in the sea?

*Tidal Power Plants*

Despite many ideas and proposals, the only form of marine hydroki-netic energy that has been developed to any degree to date is tidal power. The La Rance power station, essentially a 2,800-foot-long dam across the Rance River estuary in France, was constructed over three years and opened in 1966 (figure 10.5). The site had been identified for decades as an ideal location for an installation to harness tidal energy because of the extreme tidal range, which averages 26 feet but reaches a maximum of 44 feet, and also because of the size if the estuary. The 24 turbo generators, each nearly 18 feet in diameter, produce electricity during both flood and ebb tides and have a maximum capacity of 240 MW but an average capacity of just 57 MW. The plant paid for itself after twenty years of operation and now provides a little over one-tenth of 1 percent of France's electrical energy, quite small in the big picture of things, but the energy is clean and renewable.

For forty-four years, La Rance was the largest tidal power station in the world, in fact one of the only ones ever built. But, it was surpassed not long ago by the Lake Sihwa station in South Korea in 2010 (figure 10.6). This power plant uses a 7.5-mile-long breakwater and generates power on the incoming tide but can still produce 254 MW of power.

The Bay of Fundy between Nova Scotia and New Brunswick is generally recognized as having the largest tidal range on the planet, with a mean spring tidal range of 47.5 feet and a maximum range of 53.5 feet. These extremes require very careful assessment of where you tie up your boat or where you decide to take a hike. Because of the huge tidal range, tidal power has been considered here since the 1930s. In 1984, over fifty years later, the province of Nova Scotia commissioned and then

FIGURE 10.5. The La Rance power station along the South Atlantic coast of France was the first large-scale tidal power planet in the world. (Photo: Dani 7C3 licensed under CC-BY-SA-3.0 via Wikimedia)

completed a modest tidal plant on the Annapolis River on the Bay of Fundy. The facility generates up to 20 MW of power (recall that 1 MW can generate power for 1,000 people or more, depending on per capita usage) using a straight flow turbine.

Globally there are only seven tidal power stations on the planet with a total capacity of 522 MW, with La Rance, France, and Siwha Lake, South Korea, making up 94.6 percent of the total. One more large station is under construction in Korea, and nine more are in the planning stages in other countries. One of these, on the Swansea Bay Tidal Lagoon in the United Kingdom, would involve construction of a 6-mile-long breakwater to create a 4.4-square-mile lagoon. Tidal flow in and out of the lagoon would generate 240 MW of power using reversible turbines installed along its length. A planning application was approved for the $1.4 billion project in 2015, with some additional steps still ahead, including a critical component, securing a government subsidy.

Significant barriers to large-scale use of tidal energy remain. Not the least of the challenges are appropriate locations where large tidal ranges occur with large estuaries or coastal bodies of water with narrow

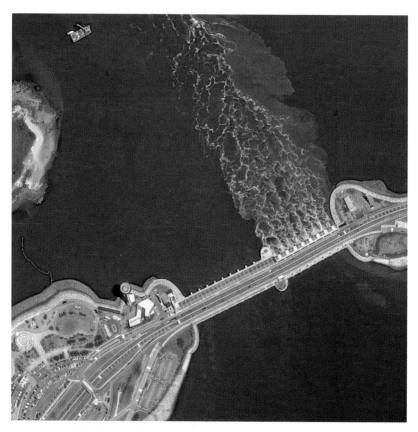

FIGURE 10.6. The Siwha Lake, South Korea, tidal plant is now the largest in the world. (Imagery 2016 CNES/Astrium, Map data © 2016 SK telecom)

entrances that can be dammed and which do not cause obstruction to boat or ship traffic or produce unacceptable environmental impacts. While tidal energy is clean and renewable, based on its limited utilization and few installations to date, it is unclear just how much tidal energy will be able to contribute to future global electricity needs.

## Hydrokinetics: Utilizing Coastal or Tidal Currents

Utilizing large, slowly rotating turbines to capture the energy contained in coastal ocean currents has been proposed for many locations, but to date challenges in the size, installation, and maintenance has made this approach too expensive. The Florida Current, which is the beginning of the Gulf Stream, is one location that would appear to be nearly ideal. It lies off the coast of Florida, just 10 miles from Miami, so is easily accessible. The current is well defined, very constant, and transports massive volumes of water (nearly 40 million cubic yards/second, which is about 137 times more water than the average flow of the Amazon River) at velocities of up to 6 feet per second (4 mph). The challenges include developing and building the huge turbines and then installing them either as a floating facility or anchored to the seafloor. The latter would be far more complicated due to water depths.

There are a number of other narrow passages in coastal waters around the world, typically between islands or between the mainland and an island, where current velocities are known to be strong, that would seem to provide good sites for pilot installations. One of these is Deception Pass, a strait between Whidbey Island and Fidalgo Island in Puget Sound, Washington. Strong currents, with velocities of up to 9 miles per hour flow through the strait on both ebb and flood tides, which leads to visible whirlpools, eddies, and standing waves.

The first commercial ocean power plant using a coastal current, which is actually the tidal flow in and out of Strangford Lough, a large bay along the east coast of Northern Ireland, was connected to the electrical grid in July 2008. The SeaGen generator, also known as a tidal stream generator, or tidal energy converter (TEC), weighs about 1,100 tons. The distinction between a tidal stream generator and a tidal power station, like La Rance, is that the latter requires complete damming of a tidal basin or estuary, whereas a tidal stream generator simply uses the tidal current. The SeaGen installation consists of what looks like two massive airplane propellers attached to a 140-foot-long horizontal beam, which is then fixed on a circular vertical column (figure 10.7). The installation produces 1.2 MW for 18 to 20 hours a day, enough to provide power to 1,500 homes. This system is being used as a prototype for larger installations, which they hope to increase to 10 MW within the next several years.

Construction of what is planned to be the world's largest tidal energy project took an important step forward in 2014, when it received initial funding to begin placement of the first four turbines on the seafloor off northeastern Scotland. The project operated by Atlantic Resource's MeyGen is planned for the Inner Sound of Scotland's Pentland Firth and would at full buildout have the potential to power nearly 175,000 homes through a network of 269 turbines generating nearly 400 MW. The first power is expected to be delivered to the national grid in late 2016.

FIGURE 10.7. The SeaGen system uses a tidal current to generate electricity. The first installation was along the coast of Northern Ireland. (Photo: © Siemens AG)

The use of streaming tidal energy is still in its infancy, although the United Kingdom has placed a priority on developing a significant portion of its future electrical power through renewable ocean resources. Projects such as these, which are in early stages of development and installation, often suffer from both time delays and increasing construction costs, so these should be expected.

## Power from Waves

Like the rise and fall of the tides, waves crashing on beaches around the world day after day, or large swells passing by ships at sea, have for decades inspired people to believe that there must be some way to harvest all of that energy. And while wave heights and periods can change rapidly, there is still a lot of energy out there. Dozens of devices have been designed, engineered, and built in ambitious efforts to capture some of that energy and turn it into electricity, but for the most part these are still in the development and trial phases. As of early 2016 there were 256 different wave energy devices that had been developed or patented by companies, academic institutions, or start-ups, all clearly trying to come up with a workable, affordable device with commercial applications. A handful of countries, including the United States, Russia, Portugal, Norway, Sweden, Scotland, India, and Japan, have built small pilot wave power plants to assess their effectiveness. At least twenty companies around the world have or are developing wave energy technologies.

The technologies that are the most advanced for the conversion of wave energy into electrical power fall into four general mechanical systems:

1. An oscillating water column, where the rise and fall of water in a tube as a wave passes forces air through a turbine that generates electricity.

2. Attenuators, such as the snakelike Pelamis, where an elongate and segmented floating tube, about the length of a submarine, flexes as waves pass by, with the up-and-down motion driving a generator.

3. Overtopping, where waves spill over into a floating reservoir with the return flow of water into the ocean driving a turbine.

4. Wave buoys or point absorbers, where passing waves cause the structure to rise and fall and uses this motion to turn a turbine.

As of 2016 there had been only a single example of a wave power system that, for a short period, provided electricity to a grid. In September 2008, Portugal completed the installation of the world's first commercial power plant that captured wave energy. It used three articulated steel "sea snakes" (the Pelamis system), 3 miles off the country's northern coast (figure 10.8). The four main cylindrical tubes that make up each unit are about 11 feet in diameter and about 400 feet long and extract power from the wave-induced motion of the hinged joints. Hydraulic rams pump high-pressure oil to hydraulic motors that drive electrical generators. Test projects offshore of the United Kingdom generated power before the first commercial installation was deployed off Portugal. The Portuguese installation produced 2.25 MW, enough to power about 1,500 homes with electricity, but only operated for four months, until technical difficulties led to the system being towed back to the shoreline. The global financial crisis made future financing difficult. The project designers subsequently tested a next generation model off the Orkney coast in 2010, and additional and larger arrays are being planned.

The potential amount of wave energy that could be utilized is significant and many different technologies have been developed, but we still have a way to go. The development of renewable ocean energy in the

FIGURE 10.8. Pelamis, off the coast of Portugal, which used a series of large, articulated floating cylinders, was the first and only wave power system that went into operation. It operated for four months in 2008. (Photo: courtesy of Wave Energy Scotland)

United States has been hampered to date by a number of bureaucratic challenges. The regulatory system was not designed to encourage pilot and demonstration projects, and, related to this lack of certainty, there has been a general lack of investment in basic research and the development of new technologies. State and federal governments need to put in place permitting processes that encourage development of demonstration projects while being sensitive to protection of the marine environment. If we are to make any progress with wave energy, regulation of ocean power development needs to be clear, efficient, and organized, with a single lead agency coupled with common sense. It is still a little early to predict with any certainty how important the role of wave energy will be in our future, but the potential is very large.

*Ocean Thermal Energy Conversion*

Ocean thermal energy conversion (OTEC) is a process that uses the differences in ocean temperatures to generate electricity through use of a

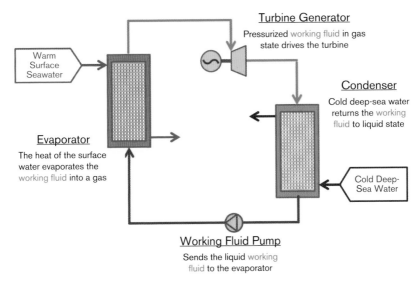

FIGURE 10.9. Flow diagram for the process involved in producing power using temperature differences in seawater, or ocean thermal energy conversion (OTEC).

heat exchanger. The ideal conditions exist when there is a temperature difference of about 36°F (20°C) between warmer surface waters and cold deep water. These conditions exist in the tropical or semitropical oceans, between about 23.5° north and south of the equator. Warm seawater passes through an evaporator and vaporizes the working fluid, ammonia (figure 10.9). The ammonia vapor flows through a turbine, which turns a generator, making electricity. The ammonia, now at a lower pressure vapor, is condensed through contact with deep, cold seawater. The liquid ammonia leaves the condenser and is pumped back to the evaporator to repeat the cycle.

There are many potential benefits of OTEC. To point out a few, it is a constant, steady, year-round power without any need for energy storage; it is a clean energy source without any emissions or by-products; and it is environmentally sustainable.

The idea of capturing the potential energy or heat exchange between cold and warm water is not a new one, but despite many years of experimentation, there is not yet any commercial development or application for electrical generation. The process requires pumping large amounts of cold water from at least a mile down and up to the surface in a large-diameter pipe, which itself uses considerable power. The United States developed what is probably the world's leading testing site on the island of Hawai'i in 1974 and has been experimenting and developing different approaches and materials ever since. While there are also test or pilot facilities operating at several other locations around the world, applications have to date been quite small, in the kilowatt rather than megawatt scale. OTEC costs remain significantly higher than other sources of energy, however, which is an unresolved challenge.

## THE FUTURE OF OCEAN RENEWABLE ENERGY: WHERE DO WE GO FROM HERE?

The coastal ocean around the planet has vast amounts of potential renewable energy, whether from wind, waves, tides, currents, or temperature differences, and these are typically adjacent to large coastal concentrations of people. A number of different technologies have been developed to harvest this energy, and more are emerging all the time. But in contrast to our present heavy global dependence on fossil fuels for generating electricity, coastal ocean energy is clean and sustainable, and once a generating system is in place, there are no fuel purchasing costs. Offshore wind energy is by far the most developed coastal ocean energy source. Northern Europe is the global leader in the construction of offshore wind farms, setting an example for the rest of the planet. Through a combination of setting national energy priorities and encouraging development and construction through government incentives and a clear and straightforward approval process, eleven European countries have been successful in installing 3,230 offshore turbines at eighty-four individual wind farms having a total capacity of 11,000 MW.

Considerable experience has now been gained, problems and impacts have been resolved or mitigated, and technologies have been refined, such that the United States should learn from Europe and move aggressively toward developing our own offshore wind energy. Approvals in state waters, such as has taken place in Rhode Island, may initially provide the clearest path forward, although the federal government can be a leader by providing a clear pathway to permitting and construction, encouraging investments through tax credits, and making the transition to renewable ocean energy a priority rather than a curiosity. All new technologies have usually required some support or subsidy to develop and implement, but we need to invest in sustainable energy sources for our long-term survival.

CHAPTER II

# Groundwater and Petroleum Withdrawal

*Subsidence and Seawater Intrusion*

Venice, Italy, has to be the planet's poster child for coastal subsidence. Images of well-dressed visitors with their pants rolled up and plastic bags tied around their legs, walking through or sitting in sidewalk cafés in St. Mark's Square with water up to their knees are vivid reminders of the low point to which Venice has sunk (figure 11.1). This is not a new process; it has been going on for centuries. This very old city was built on the subsiding delta of the Brenta River along the northeastern coast of Italy, which was really a group of small muddy islands sitting in a coastal lagoon connected by hundreds of bridges. Although the islands were very low to begin with, natural consolidation and subsidence of the thousands of feet of underlying deltaic sediments has continued over the centuries, accelerated by overpumping of groundwater from the 1930s to the 1960s. Although the pumping was halted about 40 years ago, settlement hasn't stopped. As the delta continues to subside, so does Venice, along with its historic buildings, monuments, and squares. Many ground floors are now below sea level and a number of former streets have become canals. High tides and storm surges submerge sections of Venice under as much as 3 feet of seawater, although this doesn't appear to have slowed the flow of tourists.

While the city has been fighting the problem of flooding and inundation for over five hundred years, the efforts have met with limited overall success. Multiple problems continue to plague the city, including the chemical breakdown of the marble foundations of the older buildings

FIGURE 11.1. Flooded Piazza San Marco, Venice, Italy. (Photo: London road, licensed under CC BY 2.0 via Flickr.

from submergence in seawater and discharge of runoff and untreated wastewater at high tides directly into the lagoon and canals. The most recent attempt to alleviate the problem of coastal flooding is the construction of a series of three massive floodgates or barriers at the three main entrances to the Venice Lagoon. This $6.7 billion government project, named MOSES (Modulo Sperimentale Elettromeccanico, or Experimental Electromechanical Module, in reference to Moses's biblical parting of the Red Sea) consists of seventy-eight mobile floodgates gates that are hinged and designed to be elevated at times of impending flooding, thereby protecting Venice from very high tidal or storm flooding (figure 11.2). The initial planning for the project was started in 1981, and construction began twenty-two years later, in 2003, after considerable delays and debate over the effectiveness of the massive project. Each of the seventy-eight individual gates is 92 feet long and 65 feet wide and weighs 300 tons. They are supported on the seafloor by a system of 125-foot-long steel and concrete pilings driven into the bed of the lagoon at the three entrances. Construction was slowed by ongoing corruption scandals involving contractors and politicians in the summer of 2014 but was initially scheduled for completion in 2016. After all the planning, engineering, and construction, however, while the floodgates

FIGURE 11.2. Italy is constructing a $6.7 billion system of floodgates to close off the Venice Lagoon at very high tides. (Photo: Chris 73 licensed under CC BY-SA 3.0 via Wikimedia)

gates are designed to resist a flood tide of up to nearly 10 feet, this set of massive and expensive barriers will not halt the subsidence of Venice or the increasing rate of sea-level rise.

## COASTAL SUBSIDENCE FROM FLUID WITHDRAWAL

Pumping out water or oil from subsurface sediments is the most common cause of human-induced subsidence in many coastal areas. In any fluid-saturated sediment, sand, for example, the voids or pore spaces between the individual grains are filled with water or oil in some places, and these fluids carry or support some of the overlying load of sediment and water. When the groundwater or petroleum is removed, the load previously carried by the fluid will be transferred to the individual sand grains and the sediments will undergo compaction or rearrangement of the grains. This compaction will extend through the entire thickness of sediment or soil that has had fluid removed, and there is often a direct relationship between the amount of fluid withdrawn, as measured by the drop in groundwater level, and the amount of ground surface subsidence.

### Santa Clara Valley–South San Francisco Bay

In the southern San Francisco Bay Area, in the Santa Clara Valley, a quadrupling of the rate of groundwater pumping for agricultural, domestic, and industrial uses from the 1920s up to the 1960s led to a drop in the groundwater table of almost 250 feet. Surveys revealed that for every 20 feet of groundwater table drop, the ground surface in San Jose subsided as much as a foot. The ground surface subsidence affected

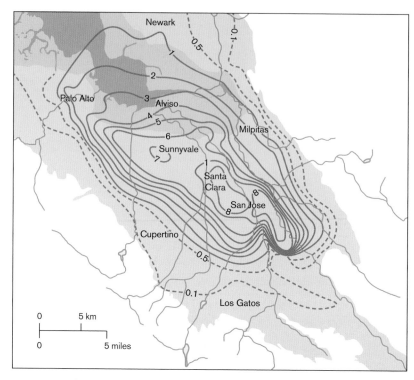

FIGURE 11.3. The city of San Jose, in the Santa Clara Valley, California, subsided as much as 8 feet between 1933 and 1967, although maximum total subsidence has reached 13 feet. Contours indicate feet of subsidence. (U.S. Geological Survey)

an area of 250 square miles and reached a maximum of 13 feet in downtown San Jose by 1967 (figure 11.3). This is more than an inconvenience. Sewage lines, street runoff, and storm drains all flow by gravity, and they don't flow uphill. The northwestern end of this area is directly adjacent to the south end of San Francisco Bay, so neighborhoods and infrastructure are now more susceptible to flooding at high tides, a situation that will only worsen with additional sea-level rise. Groundwater pumping in the Santa Clara Valley has been nearly eliminated or heavily taxed, which has allowed some recovery of the water table and a halt in the ground subsidence.

## Long Beach–Wilmington Oil Field, California

At one time, Long Beach, California, was known as the "Sinking City." The production of oil and gas from the very large Wilmington Oil Field,

FIGURE 11.4. Up to 29 feet of subsidence took place in the Long Beach–Wilmington area of Southern California from 30 years of oil extraction. The base of this fire hydrant was at ground level in this 1950 photograph. (Photo: City of Long Beach)

where 3.75 billion barrels of oil have been extracted, led to a bowl of subsidence that ultimately reached a maximum of 29 feet and was centered in and around a major international port and a large naval shipyard. While some minor subsidence from groundwater pumping was noticed in the 1920s and 1930s, it was shortly after oil and gas pumping began that serious ground settlement began to be documented. This is an intensively developed industrial area, so people began to notice when things started sinking and cracking. By 1951 the subsidence reached 2 feet per year. Wharves began to be flooded at high tide, pipelines and railroad tracks were warped and sheared; streets, bridges, and buildings were displaced and cracked, and 95 oil wells were damaged or sheared off by underground movement. Structures and utilities were all affected by the subsurface vertical and horizontal movements associated with the large depression that had affected 20 square miles by 1958 (figure 11.4).

Studies soon came to the conclusion that withdrawal of the oil from the subsurface sediments had caused compaction and led to sinking of the ground surface. The agreed-on solution, which is now routinely done for enhanced oil recovery elsewhere, was to repressurize the oil reservoir by pumping in seawater. This was a complex undertaking, but all land-ownership, responsibility, and cost issues were finally resolved and fluid injection began, at rates of about one million barrels of seawater a day. This halted the subsidence, and because oil is lighter than water, the water injection also caused the remaining oil to rise in the subsurface reservoirs, and additional oil could be extracted. Damages and remediation costs ultimately reached an estimated cost of $100 million.

*Houston-Galveston Area, Texas*

Perhaps more than any other large metropolitan area in the United States, the Houston area has been negatively affected by widespread subsidence, due primarily to groundwater pumping but also from oil and gas extraction. Land subsidence was first noticed over a century ago, in the early 1900s, in areas where groundwater, oil, and gas were all being pumped out. As groundwater extraction continued throughout the twentieth century, patterns of subsidence in the greater Houston area closely followed the temporal and spatial patterns of subsurface fluid withdrawal.

While subsidence on a regional basis may be difficult to detect initially, there are areas in and around Houston where the impacts are quite clear. Because of the low elevations and direct connections to Galveston Bay and the Gulf of Mexico, up to 10 feet of subsidence has shifted the location of the shoreline. This increased the frequency of coastal flooding and caused major damage to transportation and industrial infrastructure and led to major expenditures in levees and other protective structures as well as loss of wetlands.

The impacts of subsidence are well illustrated by the history of the Brownwood subdivision of Baytown, built in the 1930s as an upper-income development along the edge of Galveston Bay. Even when constructed, the subdivision was only 10 feet above sea level at most. But by 1978 over 8 feet of subsidence had taken place (figure 11.5), leaving Brownwood just a drowned reminder of the hazards of coastal subsidence from fluid withdrawal. Activation and surface displacement (creep) along faults in the Houston-Galveston area were observed as an additional impact of ground settlement. Monitoring of fault creep, water

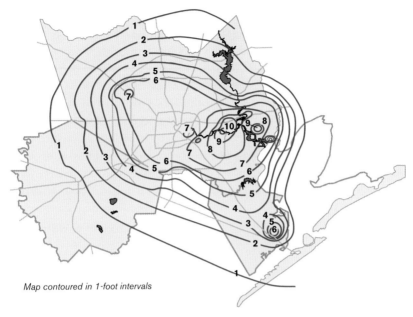

Map contoured in 1-foot intervals

FIGURE 11.5. Ground subsidence in the greater Houston area has reached a maximum of 10 feet. (Courtesy: Houston-Galveston Subsidence District)

levels, and land subsidence demonstrated a clear cause-and-effect relation, with movement along faults caused by water-level decline and the accompanying subsidence.

As the fate of Brownwood and other impacts of subsidence were gradually recognized, public awareness increased and actions were initiated to halt the subsidence. A reduction in groundwater extraction was seen as the only long-term solution, and in 1975 the state legislature passed a bill establishing a Subsidence District, the first of its kind in the United States, which had the power to restrict groundwater pumping. Those reductions, combined with the importation of surface water from a reservoir constructed on the Trinity River, soon slowed water-level decline and ground subsidence, and also halted or reduced fault creep. While conversion of water source from groundwater to surface water has greatly alleviated the problem in the Houston–Galveston Bay area, subsidence is accelerating to the west, where groundwater use has increased. Active subsidence has shifted from the low-lying, tide-affected areas toward higher elevations inland.

Economic costs or damage from long-term subsidence in the greater Houston-Galveston area were large, but determining the actual economic cost of subsidence is hard to quantify, and most published estimates are necessarily vague. One source in 1983 stated that "many millions of dollars" were spent elevating structures such as buildings, wharves, and roadways and constructing levees to protect against tidal inundation, and also that "millions of dollars" were spent on repairing damage due to fault movement. For the period from 1969 to 1974, the average annual cost to property owners was estimated at about $133 million (in 2015 dollars). Damage and repairs to industrial facilities are among the greatest costs, including relocating dock facilities, constructing hurricane levees, and repairing drainage problems at refineries along the Houston Ship Channel. Estimated total cost for just two refineries was about $500 million in 2015 dollars. Using these estimates as guides, it appears reasonable to conclude that the overall economic impacts of subsidence in the Houston-Galveston region are at least several billions of dollars.

## New Orleans

By any measure, New Orleans is an interesting and vibrant city. With its history, architecture, culture, music, and food, the city has always drawn people from throughout the country, and around the world for that matter. The city's geological setting, however, also sets it apart but not in any positive way. New Orleans developed on one lobe of the complex of deltaic sediments deposited by the Mississippi River over thousands of years. The Mississippi River drains the largest watershed in the United States, 1.1 million square miles, or just over 40 percent of the entire area of the contiguous United States, including all or parts of thirty-one states and two Canadian provinces. The river delivers about 150 million tons of sediment annually, on average, to the Gulf of Mexico. Some of this has gone into visible portions of the delta, but much of the finer-grained sediment is carried out into the Gulf to build the massive underwater portion of the delta. A dynamic equilibrium existed, prior to any human intervention, between distributary channel pathway, sedimentation, and subsidence.

Soon after the earliest settlers arrived in the New Orleans area, they began to alter the natural system with the construction of levees. By the early 1900s, the Army Corps of Engineers had stepped in to take over the management of the river and its distributaries by building levees to contain the river and its course to the Gulf of Mexico. The river engineering, as is often the case, upset the natural balance by eliminating the sediment

FIGURE 11.6. Flooding of New Orleans during Hurricane Katrina as levees and floodwalls failed. (Photo: Lieutenant Commander Mark Moran, NOAA Corps, NMAO/ AOC. Licensed under CC BY 2.0 via Flickr)

deposition on the terrestrial portion of the delta, at New Orleans, for example, which continued to subside as there was no new sediment to keep pace with the sinking. Disruption of the natural sediment supply, combined with groundwater, oil, and gas extraction, has led to continued subsidence of New Orleans and the surrounding areas, most of which now sit as much as 10 feet below the level of the Mississippi River.

The New Orleans metropolitan area had a population of about 1.3 million in August 2005 when Hurricane Katrina hit. Most of these people lived below sea level, and 80 percent of the city was flooded when the levee system failed (figure 11.6). The storm surge had heights ranging from 7 to as much as 16 feet, and in the city the surge led to twenty-three breaches in floodwalls and navigational canal levees. Katrina was the costliest natural disaster to ever strike North America and has been called the worst engineering disaster in U.S. history. Total damages might reach $200 billion, and although the official death toll was 1,500, many people were never accounted for. Rebuilding the levee system to withstand a Category 5 storm is projected to cost $32 billion, or $66,000 for each of the original 485,000 residents. Hurricanes have hit

New Orleans in the past, and they will without question hit again. With tens of thousands of residents living below sea level on a subsiding delta, protected by artificial levees, and with a rising sea level, the prognosis is not bright.

## Groundwater Overdraft and Seawater Intrusion

Somewhere around 40 percent of the world's population (about 3 billion people) live within 100 miles of a coast, and the twenty largest metropolitan areas sited on coasts are home to about 250 million people. Many of these coastal areas have relied on groundwater throughout much of their histories for agricultural, domestic, and industrial uses. As increasing numbers of people have migrated to coastal zones, the demand on groundwater has increased, and in many areas the aquifers have been overdrafted. Simply put, more water is being pumped out over time than what is naturally seeping back in. The response to this imbalance is easy to understand but not always easy to see.

The difference between how much water is removed from an aquifer each year (and this can be from natural discharge at springs or into streams or through pumping from wells) and the amount that seeps back in (through the percolation of rainwater through the soil or through the bottom of riverbeds, lakes, or percolation ponds) will determine how the water table or the water level in the aquifer will change over time.

In many coastal areas around the world, freshwater aquifers are in direct contact with the ocean along the offshore seafloor. Prior to heavy human use of these aquifers, the groundwater table beneath the surrounding hills and alluvial valleys or river deltas would slope gradually downhill toward sea level. There was a balance or equilibrium, with the pressure of the freshwater flowing toward the ocean balanced by the pressure of the seawater directed landward. Under these conditions, the pressures of each of these two water bodies balanced one another. Over time, as water demands increased, more wells were drilled and more water was pumped out than was being naturally recharged. This always leads to the same result: the water table drops. With a lower water table, the pressure from the freshwater on the landward side can no longer resist the pressure of the seawater from the ocean, and salt water begins to intrude into the aquifer. If overpumping continues, the groundwater table will drop deeper, allowing a wedge of salt water to intrude farther inland. Brackish or salt water will begin to appear in wells, and none of the users of freshwater, whether agricultural, domestic, or industrial, are happy with salt water. Just one-third of a teaspoon of salt will make the water in a one-gallon container too salty for safe drinking or most uses.

Saltwater intrusion was noticed as far back as 1845 on New York's Long Island, and it is now a problem in coastal aquifers around the world. Since the 1940s saltwater intrusion has forced the abandonment of more than one hundred wells on the Cape May Peninsula, New Jersey, a popular vacation destination. As popularity of the resort grew, more people came and used more and more water from the community's wells. With a dropping water table, seawater intruded farther into the aquifer, and more and more wells became contaminated with salt water. The solution at Cape May was the construction of a $5 million desalination plant to remove the salt from the brackish well water. This allowed the residents to continue to use now treated brackish water and decrease pumping so as to reduce the rate of inland migration of the saltwater wedge.

Virtually every state along the Atlantic and Gulf coasts must deal with the problem of saltwater intrusion. Savannah, Miami, and Tampa Bay are just three large cities that have been dealing with saltwater intrusion for years. Ninety percent of the water supply of South Florida comes from underground aquifers, chiefly the Biscayne Aquifer. With greater depletion of the aquifer and recent droughts, there hasn't been enough rainfall to replenish the groundwater. The natural response is for salt water to advance farther inland. Miami's earliest problems stemmed from inadequately controlled tidal drainage canals, which reduced freshwater levels and provided easy access for seawater to move inland during dry periods. Placement of tide gates in the canals helped retard and reverse the intrusion. But today, Hallandale Beach, Pompano Beach, Dania Beach, Lantana, and Lake Worth are all experiencing their own saltwater problems, and cities a little farther inland, from Jupiter to Florida City, like Fort Lauderdale, Hollywood, and Miami, are increasingly threatened.

Most of Florida suffers from a much larger geologic condition, however, which not only affects saltwater intrusion but also the impact of sea-level rise on the shoreline of the state. Most of Florida is underlain by limestone, which dissolves in groundwater, and therefore is riddled with solution pits, caves, and sinkholes, making the rock like Swiss cheese. These interconnected cavities are often exposed or open on the seafloor, which allows salt water to penetrate inland more directly and also makes a rising sea level much more problematic for a place like Miami. In 2016 there were 417 condominium towers—a total of 50,060 units—under construction from Miami to West Palm Beach, and not one of those towers was designed taking sea-level rise into account.

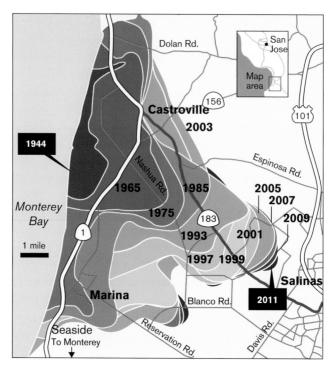

FIGURE 11.7. Seawater intrusion in central Monterey Bay has extended nearly 8 miles inland to the city of Salinas.

Seawalls or other barriers can be built to hold back the rising ocean, at least for a while, but the salt water can flow in under the walls and penetrate inland, causing any of a number of problems.

Continuous heavy pumping of groundwater from the lower Salinas Valley along the Monterey Bay coastline of Central California since the early 1940s, primarily for a $4 billion agricultural industry, has depleted the groundwater storage and dropped the water table. Intruded salt water led to reduced crop yields and increased costs of drilling deeper wells or new wells farther inland to replace wells yielding saline water. Over one hundred wells were capped or abandoned in one small area near Castroville between 1943 and 1968 because of excess salt. Groundwater accounts for over 95 percent of all the water used in the Salinas Valley for both agricultural and growing domestic needs, but the increased pumping led to the intrusion of salt water in the 180-foot aquifer up to 8 miles inland to within a half mile of the city of Salinas (figure 11.7).

After sixty years of dedicated effort, the Salinas Valley farmers have gotten the problem nearly under control and provide an example of what can be done with integrated approaches to groundwater management. Two reservoirs constructed in the 1960s in the upstream portion of the watershed allowed for water storage for recharging the groundwater during the periods of peak pumping. A wastewater treatment plant in Castroville now produces water clean enough for crop irrigation so farmers can shut down their salty wells. An inflatable rubber dam on the lower Salinas River allows for diversion of water that would have flowed to the sea to be used for crops. One of the most important changes has been the slow conversion from sprinkler and furrow irrigation to drip irrigation, and also installation of high-tech soil moisture sensors and flow meters and the use of land leveling. While year-to-year variations in rainfall make annual water use difficult to compare, groundwater pumping peaked in 1997 at 552,000 acre-feet, and in 2011 this had fallen 27 percent, to 404,000 acre-feet. (An acre-foot is a common unit for discussing large volumes of water and is an acre—about the size of a soccer or football field, 43,560 square feet—covered with one foot of water, or 325,851 gallons, roughly what one or two families would use in a year.) Seawater intrusion has been significantly slowed, and groundwater tables are slowly rising.

The coastlines of many European countries also face problems of saltwater intrusion into heavily used coastal aquifers, including Turkey, Italy, France, Spain, Portugal, Denmark, and Germany. These are not just localized issues. Recent studies in Spain indicate that 60 percent of the nation's coastal aquifers are contaminated by saltwater intrusion, including some of the most important systems in terms of economic impacts.

Since the early 1960s China's coastal aquifers have been monitored for seawater intrusion. Coastal aquifers underlie hundreds of square miles, the land is intensively farmed and irrigated, and millions of people depend on the success of the harvest, making saltwater intrusion a significant concern.

## LOOKING TO THE FUTURE: WHERE DO WE GO FROM HERE?

Looking into the future, climate and land use changes will also have significant effects on the problem of seawater intrusion through their impacts on the hydrologic cycle. A warmer planet and more people will mean greater water requirements, much of this coming from groundwater. The amount of precipitation is ultimately the source of all groundwater, and as precipitation patterns change spatially and temporally around the Earth, this will affect the amount of water entering the subsurface and recharging the water table. Altering vegetation patterns through deforestation, or other land use changes, can affect the amount of *evapotranspiration* (the amount of precipitation that is evaporated from the ground surface or water bodies and the water used by plants that is ultimately lost through leaves to the atmosphere), how much water is retained and can percolate into the ground, and how much will become surface runoff.

By the time a farmer's crop dies from irrigating with salt or brackish water, however, it's a little late; so today most coastal water wells are or can be monitored regularly by electrical conductivity analysis, which provides an accurate measure of salinity. Combined with measures like those used by farmers in the Salinas Valley—soil moisture sensors and moving from overhead sprinkler or furrow irrigation to drip—this can greatly reduce water use and therefore the demand on groundwater aquifers, allowing water tables to rise and freshwater aquifers to come into balance with seawater again. These and other measures that have proven effective at reducing or mitigating the problem of seawater intrusion need to be employed in coastal areas around the world where intrusion has become a problem. For too long water has been used as if it were an unlimited resource. The use of highly inefficient flood or furrow irrigation or overhead sprinklers no longer makes sense in any farming operation. Recent legislation passed in California for the first time requires that all groundwater basins prepare management plans such that each basin comes into balance with the amount of water withdrawn being no greater than the amount recharged or entering the aquifer. Overdrafting will gradually be halted.

# CHAPTER 12

# Desalination

*Freshwater from the Sea*

## GLOBAL DISTRIBUTION OF WATER

Desalting seawater seems like a logical solution to our global water shortages, simply because all but 3 percent of all the water on the planet is in the oceans. The natural process of evaporation from the ocean desalts about 35 billion acre-feet of seawater every year and through the processes of condensation and precipitation makes that freshwater available—and it doesn't cost us anything. This huge volume of freshwater is enough to provide 4.7 acre-feet (1.5 million gallons) for every one of the 7.4 billion people on the planet yearly, which is about four times the amount needed by a person for all purposes, in addition to taking care of environmental requirements. Unfortunately, all of that precipitation isn't readily accessible to the people of the planet, or the landscape, simply because a very large part of it is stored as ice sheets and glaciers or under the ground; only about 0.3 percent of all the freshwater on the planet sits at the surface in lakes and rivers.

Globally, 8 percent of all water use is for domestic purposes, 22 percent is used for industry, and 70 percent is used for agriculture. While most developed countries have water supply and treatment systems, there are still about one billion people on Earth who do not have access to improved drinking water sources, and as a result, 3,900 children die every day from dirty water or poor hygiene.

As Earth's climate continues to warm, the amount of easily accessible freshwater found in rivers, lakes, and reservoirs will likely decline. In

FIGURE 12.1. Bristlecone pines can live up to 5,000 years. (Photo: Rick Goldwaser licensed under the Creative Commons Attribution 2.0 via Flickr)

times of severe drought, it declines a lot. Dendrochronology, or the study of tree rings, allows us to look back as far as five thousand years in the American Southwest and determine when the area experienced prolonged periods of drought and how long they lasted (figure 12.1). When water is plentiful, trees grow wider rings; but when it is dry, growth slows and the rings are thinner (figure 12.2).

What does this history tell us about what we might expect in the future? Studies of ancient climate suggest that the water crisis of the past decade, and even the 1930s Dust Bowl, pale in comparison to a series of droughts that struck the southwestern United States 700 to 1,000 years ago. The periods 1020–50 (30 years), 1130–80 (50 years), and 1276–99 (23 years) were exceptionally dry and were preserved as very narrow or slow-growth rings in bristlecone pines. These extended droughts have been confirmed with other lines of evidence as well. The American Southwest has a long history of prolonged droughts, and similar conditions should be expected in the future.

Unfortunately for us, freshwater isn't always where people are, although such a connection was sold as gospel ("rain follows the plow") and attracted the first farmers to the arid Midwest and West of the United States. Those geographic amenities that have drawn humans

FIGURE 12.2. Dendrochronology, or the study of tree rings, can extend our climate record back hundreds to several thousand years. (Photo: Courtesy LTRR [Laboratory of Tree Ring Reseach], University of Arizona)

to the coastline over the centuries (e.g., adjacency to the ocean, pleasant climate, flat fertile deltaic or coastal plain land for agriculture) aren't the same attributes that bring large amounts of rain. In many coastal regions around the world, expanding agriculture and growing cities have tried to have their cake and eat it too. Government at all levels believed it was their responsibility to deliver water to the residents and went after much of the unclaimed water (as well as some of the already claimed water, regardless of existing users) and developed elaborate plumbing systems. In Southern California in the 1920s, a shooting war broke out between the residents of Owens Valley and the water purveyors of the Los Angeles area (the Owens Valley Wars) who built a 200-mile aqueduct to transport the water to a growing and thirsty Los Angeles. Two thousand years earlier, the Romans developed a sophisticated system of aqueducts and canals to move water from where it was plentiful to where the people were. They were excellent engineers and builders, and some of those aqueducts are still in existence today.

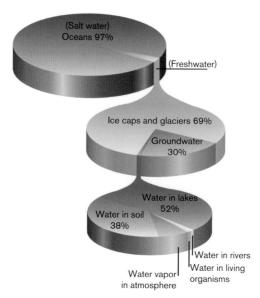

FIGURE 12.3. The percentage of the planet's total water present in each of the Earth's water reservoirs.

It appears that throughout human history it may have been water availability or scarcity, as much as any other single factor, that led to the decline of many early civilizations. Today our water consumption rate is doubling every twenty years, twice as fast as the population is growing. About a billion people lived in water-scarce areas in 2015, and this is projected to rise to 1.8 billion by 2025. By 2030 nearly half of the world's population may live in areas without adequate water supplies.

About 321 million cubic miles, or 97 percent of all the water on Earth, is salty and resides in the oceans. The remaining 3 percent is fresh, although most of that is not readily accessible. Roughly 69% of the 3 percent that is freshwater is locked up in ice sheets and glaciers. An additional 30% lies beneath the surface as groundwater, and just 0.9 percent is surface water, lakes, rivers, and swamps (figure 12.3). So it is this very small percentage of fresh surface water, and also groundwater, that provides for humanity's needs across the globe.

These percentages do change slightly from glacial periods (ice ages) to interglacial periods. During the last ice age about 10 million cubic miles of seawater was transferred from the oceans to the continents as ice sheets and glaciers. There is a lot of freshwater in the form of ice out there, but very little of it is close to where most of the planet's people live.

While river, lake, reservoir, and groundwater levels drop during extended droughts, the level of the ocean remains nearly the same. As global climate continues to warm, the freshwater locked up as ice is being reduced, with meltwater flowing into the oceans, gradually raising their level and producing more salt water. And for the first fifteen years of the twenty-first century, that melting added about 275 cubic miles of formerly freshwater to the oceans every year. This transfer will continue for centuries into the future, and very likely at an increasing rate.

## DEVELOPMENT OF DESALTING TECHNOLOGIES

The vast amount of water in the oceans combined with the frequent droughts in coastal areas where people are concentrated, the increasing per capita water usage, the present era of global warming, and the increase in overall water consumption are all reasons that people have looked to the sea for their water needs over the past century or so.

In historic times, seawater was evaporated for the salt it left behind, which was a more valuable commodity than the water itself. As people gradually spread out across the globe and began to colonize areas without adequate volumes of freshwater, particularly islands in warm or arid regions, necessity pushed the development of solar stills. These are still used as part of the survival packages stowed on lifeboats and life rafts in the event that a boat or a ship goes down. The ability to be able to produce even minimal amounts of freshwater has saved many sailors' lives.

Solar still technology is pretty simple. Using a small pan or container of seawater covered by a tilted piece of glass or clear plastic, the salt water will be evaporated, condense against the glass plate, and trickle down into a trough and then into a storage container. When seawater evaporates, the salt stays behind, as the freshwater turns to a vapor. It is a very effective mechanism for producing modest quantities of freshwater, enough to provide for the minimum daily needs in some situations. Small solar stills were developed for a single family's household needs on many small islands and proved effective, but they were not ever built to provide large volumes of freshwater.

Many ships have been supplied with distillation systems, which turn out to be a more practical approach than trying to store large quantities of freshwater. Navy vessels and cruise ships, in particular, with large numbers of crew and passengers, are almost always fitted with these systems. As seawater is used to cool the massive engines that power these ships, the seawater is warmed; with a little added heat thrown in, it will evaporate and the freshwater that condenses can be collected and stored for on-board usage.

The small Caribbean island of Curaçao was home to the first community-scale desalination plants, which were built in 1928. The Middle East followed a decade later, with the first major desalination facility constructed in 1938 in what was later to become the nation of Saudi Arabia. During World War II and later, the United States invested significantly in developing desalination technology, for both naval vessels at sea and the military deployed in water-scarce regions. By the twenty-first century, the U.S. government had invested nearly $2 billion in desalination research and technology.

Several different methods exist to remove salt from seawater or brackish water, and each of these processes takes place naturally in the oceans as well.

## Freezing

Freezing is an effective natural process for separating salt from water, because as ice crystals form from seawater they exclude salt, and therefore when the ice melts it produces essentially fresh water. The waters surrounding the ice shelves of Antarctica, for example, are essentially freshwater that is formed when the ice melts. The process works well in nature in the Antarctic and Arctic, but in a commercial operation, there is still the challenge of separating the ice crystals from the remaining salt water. A small number of plants using freezing were built for demonstration purposes in the past fifty years, but processing the ice and water mixtures has proven to be problematic and too costly for commercial use.

## Distillation

About 40 percent of the desalted water on the planet historically was produced through the distillation process, where seawater or brackish water is boiled and converted to steam or water vapor, which is then condensed to liquid water again. There are several different distillation processes that have been utilized, but each of these requires large amounts of heat. By reducing the pressure, however, water will boil at lower temperatures, which can reduce the heat or energy required. Locating a distillation plant adjacent to a large coastal thermoelectric power plant, where large volumes of very hot water are produced in the cooling cycle, can significantly reduce the energy required because the seawater is already close to the boiling point.

## Reverse Osmosis

Semipermeable membranes, which essentially work like the walls of living cells, began to be widely used in the 1970s to desalt water. Through the process of osmosis, various membranes proved to be effective in removing salt from brackish or seawater and had a number of advantages as well. Pressure is applied to force the salty water through a semipermeable membrane, which leaves the salts and other impurities behind (figure 12.4). Higher-salinity water requires more energy to push it through the membranes, but these membranes can also remove many organic contaminants and microorganisms. The required energy is mainly a function of the salt concentration in the source water. About half of the desalination plants around the world are now utilizing reverse osmosis, and this percentage continues to grow.

Desalination technology continues to improve as demand has grown, competition has expanded, and the need to reduce costs and energy has increased. There have also been a number of specific recent improvements that have been identified to increase plant efficiency, lower production costs, and improve discharge water quality. Improved pretreatment or screening of the feed water reduces the need to use chemicals that end up in the disposed brine. There are areas where the reverse osmosis process can be improved: more advanced membranes that have longer life spans can increase the amount of water that passes through the membranes, can reduce the biological fouling or growth on the membranes, and require less energy.

## Desalination Today

As of 2015, there were over 120 countries worldwide using desalination to produce freshwater, including a number of Middle Eastern nations (Saudi Arabia, Oman, UAE, and Israel), Mediterranean nations (Greece, Portugal, Spain, Italy, Gibraltar, Cypress, and Malta), and also China, Japan, Australia, and the United States. The approximately 18,400 plants operating around the world have a total capacity of about 23 billion gallons per day (bgd) of freshwater (70,330 acre-feet), which is an increase of 10% from just two years earlier. To be clear, however, a large fraction of these plants are very small facilities. Desalination

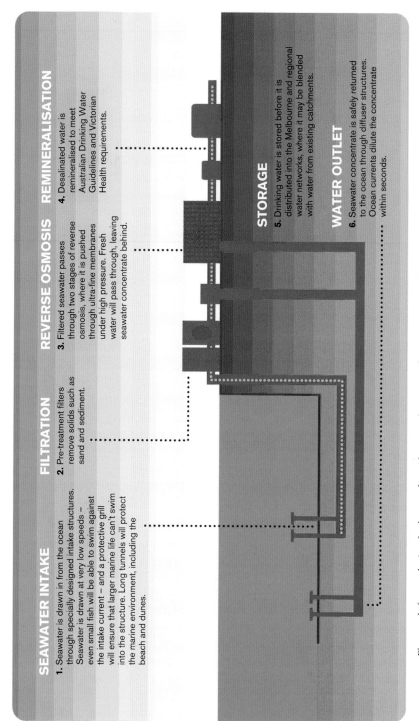

**SEAWATER INTAKE**

**1.** Seawater is drawn in from the ocean through specially designed intake structures. Seawater is drawn at very low speeds – even small fish will be able to swim against the intake current – and a protective grill will ensure that larger marine life can't swim into the structure. Long tunnels will protect the marine environment, including the beach and dunes.

**FILTRATION**

**2.** Pre-treatment filters remove solids such as sand and sediment.

**REVERSE OSMOSIS**

**3.** Filtered seawater passes through two stages of reverse osmosis, where it is pushed through ultra-fine membranes under high pressure. Fresh water will pass through, leaving seawater concentrate behind.

**REMINERALISATION**

**4.** Desalinated water is remineralised to meet Australian Drinking Water Guidelines and Victorian Health requirements.

**STORAGE**

**5.** Drinking water is stored before it is distributed into the Melbourne and regional water networks, where it may be blended with water from existing catchments.

**WATER OUTLET**

**6.** Seawater concentrate is safely returned to the ocean through diffuser structures. Ocean currents dilute the concentrate within seconds.

FIGURE 12.4. Flow path for producing freshwater from the ocean by means of reverse osmosis (from a plant in Melbourne, Australia).

does provide a significant portion of the water in the arid Middle East, with Israel producing about 40 percent of its entire domestic water need from brackish or salt water. The Ras Al-Khria plant in Saudi Arabia appears to be the largest in the world as of 2014 and has a capacity of 271 million gallons per day (mgd).

Globally desalination provides water for about 300 million people, or about 4 percent of the world's population. In volume, however, this is only about 0.5 percent of the world's total freshwater use. Over half of the feed water is from the ocean, about a quarter from brackish sources, and the remainder from rivers or wastewater. Half of the global desalination capacity is in the arid Middle East–North Africa–Mediterranean region, with Saudi Arabia responsible for 17 percent of global production. The United Arab Emirates produces 13 percent; Spain, 6 percent; Kuwait, 5 percent; and Libya, 2 percent. Most of the Middle Eastern countries also have large oil reserves, so the power production needed to run these facilities has not been a big concern. Saudi Arabia, for example, uses about one of every 30 barrels of oil they pump out of the ground for desalination, which provides about 70 percent of their water needs.

## Desalination in the United States

Over two thousand desalination plants had been built or were under contract in the United States by 2005, with Florida having the most plants, followed by California and Texas. At that time their total capacity was about 1.6 bgd, which was only about 0.4 percent of the water use in the United States. The total volume of freshwater produced by desalination, however, grew 30 percent between 2000 and 2005. Up-to-date numbers on capacity are hard to come by, and many of the values reported for the number of plants or capacity include plants that were contracted but never built, built but never opened, opened but then shut down, or small test or pilot facilities. Many of the plants listed or included in summaries of desalination capacity are small plants, often providing very high quality for specific industrial uses rather than for municipal usage.

The desalination enterprise in the United States is different from the global picture in that the dominant source of feed water globally is seawater, whereas in the United States it tends to be brackish water (51 percent), followed by river water (26 percent), wastewater (9 percent), and seawater (only 7 percent). Reverse osmosis is the most common

FIGURE 12.5. Pumps and piping at the Poseidon Carlsbad Desalination Plant, the largest such facility built to date in the United States. (Photo: George J. Janczyn licensed under CC SA 4.0 International Iicense)

process used globally and also in the United States, where about 69 percent of the desalination plants rely on this technology.

In the United States and elsewhere around the planet, interest in desalination and the number of plants proposed and under construction have increased significantly in recent years. Reasons include coastal population growth and per capita use of water, shortage of local or regional water supplies, and continuing or recurring drought conditions, to name some of the more pressing concerns.

After many years of planning, design, environmental impact assessment and review, and finally political approvals, the Poseidon Resources desalination plant in Carlsbad, California, was completed and went on line in December 2015 (figure 12.5). This is the largest, most technologically advanced and energy-efficient desalination facility in the western hemisphere and taps the world's largest reservoir of water, the Pacific Ocean. The plant has the capacity to produce 50 million gallons of freshwater every day, which is about 7 percent of San Diego County's total water usage. Although construction costs were initially estimated at $270 million, in the fifteen years from inception to completion the total price tag escalated to approximately $1 billion. There was also considerable opposition along the way as desalination rarely meets with complete public approval. In part this is because any large coastal construction project isn't usually welcomed by everyone, in part because there were specific environmental concerns that were raised, and in part because there wasn't a long track record in California, or anywhere in the United States for that matter, on the environmental impacts of desalination plants.

## THE COSTS AND BENEFITS OF DESALINATION

The complex issues surrounding desalination can perhaps be summed up this way. The ocean contains most of the world's water and is a very reliable source for coastal regions, which is independent of drought, but it is still one of the most expensive sources and comes with energy and environmental costs. The three major issues that have usually been raised and argued with every new desalination plant proposed are costs, energy requirements, and potential ocean impacts.

### Costs of Desalinated Water

The recently completed Poseidon Resources desalination plant was made possible by the commitment of the San Diego County Water Authority to purchase its entire freshwater output at a guaranteed price for thirty years. Because this is the largest and newest desalination plant in the United States, the costs of producing water should be some of the best available for comparison with other water sources.

The costs of water from the Poseidon plant are in the range of $2,100 to $2,300 per acre-foot (325,851 gallons), plus future inflation. Clearly, however, there are different families with different consumption practices living in different climates, so usage can vary, which is why calculations are based on the annual usage of one or two families (one acre-foot). The San Diego Water Authority pays about $923 per acre-foot of imported water as of 2016, so the price of desalinated water would be 2.2 to 2.4 times higher.

The price tag of this thirty-year contract for San Diego County is at least $110 million a year, which would increase the average home owner's water bill by about $5 to $7 a month. A desalination plant costing $1 billion, or annual amortization costs of $110 million, sounds like a very high price, but $5 to $7 a month, the price of two cups of gourmet coffee, doesn't sound like much at all for a family's guaranteed monthly water needs, which are not dependent on the vagaries of rainfall or allocation by some political process. So it is important to look at the water economics carefully and compare it to other daily or monthly costs, phone, Internet, and cable TV, for example. For those who buy those cute little plastic water bottles, there is another comparison to keep in mind. At a low price of $1.00 per pint for most bottled water, this translates to $8 per gallon, which is over a thousand times more expensive than high-quality distilled water from a desalination plant. If gasoline cost as much as bottled water, you would spend $100 or more every time you filled up your tank.

In addition, the actual costs of providing water from a desalination plant are not always easily compared to existing water costs or other options. Estimates or projected costs are often given for a variety of different units (acre-feet, millions of gallons, cubic meters, etc.) at different times and at different points in the supply process (e.g., at the plant; at your home or faucet) and using different interest rates, energy costs, amortization periods, and any subsidies.

It is also important to keep in mind when making comparisons between some new water source, desalination, for example, and present water prices that most cities or communities paid off their existing systems (dams, reservoirs, pipelines, treatment plants, etc.) years or even decades ago. Thus in many arid regions or states we already obtained all the easily accessible or inexpensive water and have paid off those construction costs. In addition, much of the cost of water development in the American West was federally funded (or subsidized by taxpayers nationwide). No matter what we do next, whatever source we go after, it is going to cost significantly more than what we are paying today. It is more realistic to compare the costs of desalination with the costs of some other new source of water rather than the existing costs. There is no such thing as a free glass of water any longer.

*Energy Usage and Costs*

The price of energy is the single largest variable in the costs of operating a desalination plant and can range from about one-third to one-half of the total cost (figure 12.6). Electrical costs for reverse osmosis plants average about 44 percent of the total operational costs, with fixed costs (amortization costs of the original capital investment or construction) averaging about 37 percent, and the rest being labor, maintenance, and membrane replacement. Thermal or distillation plants have even higher energy costs, usually about 60 percent of total operating costs.

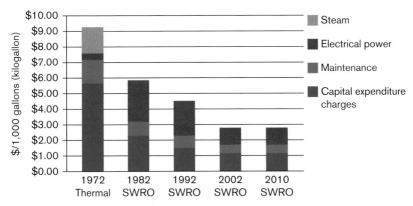

FIGURE 12.6. The cost of producing freshwater through desalination continues to decline. Evaporation was the major process used in 1972 but has been replaced by reverse osmosis (SWRO). The relative costs of individual components of the process have also changed.

Fluctuating energy prices can significantly affect the costs of desalination. An increase in the price of energy of 25 percent over time, for example, would translate into an 11 percent and 15 percent increase in water production costs for reverse osmosis and distillation plants, respectively. While desalination plants may become more efficient in the future and require less energy, for the near term there will be a direct increase in the cost of producing water with any increase in energy costs. As long as power is provided by conventional fossil fuel–generated electricity, the variations in the price of oil, natural gas, or coal will be key factors affecting the price of water. There are also economies of scale involved in desalination as in many other industrial processes. Other things being equal, increasing the size of a desalination plant will reduce the cost per gallon or acre-foot of produced water, with greater cost reductions in moving from a very small plant to a moderate-size plant.

For a comparison of the energy requirements of operating a moderate size desalination plant, a recent evaluation of the power requirements of a proposed 2 mgd desalination facility proposed for the Central Coast of California was about 6,800 MWh/yr, which is equivalent to the electrical use of a moderate-size hospital (6,600 MWh/yr) or a large indoor shopping mall (6200 MWh/yr), 50 percent more than the energy used by the existing water treatment plant (4,400 MWh/yr) and about half of the energy used by the wastewater treatment plant (12,000 MWh/yr).

There are other cost considerations as well, a major one being the period of time over which the capital cost is amortized, just like a typical home mortgage. Another variable is the experience of the project team. With more projects constructed, a team or company will do a better job of estimating total costs so that the costs of producing water in the future can be more accurately estimated. Environmental costs or mitigation measures can be an additional factor, depending on the location of the plant and local concerns or issues.

It is important to use caution in evaluating different costs for desalting water, simply because of the often large uncertainties in estimating all of the costs (energy requirements being one good example) well before a plant is even built. In spite of these uncertainties, however, the trend over time has been for desalination costs to decline (see figure 12.6). This is in part due to declining energy costs in recent decades but also to improved technologies and materials, the economies of scale when building larger-capacity plants, and savings realized with more experience in construction and operation of plants.

Many urban water users expect water to emerge from their tap when they turn it on and are willing to pay for that reliability. In arid or semi-arid regions, and Texas, Florida, and California generally fall into that category, where traditional sources of water (surface or ground) become less reliable during periods of extended drought or below average rainfall, the ability to consistently desalt seawater, regardless of the uncertainties of the weather, has a high value. The advantage of having local control over the water supply is also a benefit for a desalination facility. Local control is not possible when connected to a large water supply system, which is the case for California, where statewide decisions are made on rural/agricultural allocation of water versus urban uses.

## ENVIRONMENTAL IMPACTS OF DESALINATION

Any large industrial facility or process will have some environmental impacts. Desalination is no exception, and the potential impacts that need to be understood and mitigated to the degree possible include plant siting, construction and operation, and importantly, the effects of both pumping large volumes of salty or brackish water from a bay, estuary, or the ocean, and then any impacts of discharging the salty brine back to the ocean.

### Seawater Intake: Impingement and Entrainment

In addition to the seawater itself, the water that is pumped from the ocean or some brackish body of water into a desalination plant may contain a variety of organisms ranging in size from marine mammals and fish down to microscopic plankton. The types of organisms and their abundance will vary from place to place, depending upon the source, the time of year, and also with the depth from which water is pumped if taken directly from the coastal ocean. Any intake pipe is normally covered with a screen, and the mesh or size of the openings in the screen will determine the organisms that might be caught on the screen (impingement), and those that will pass through and be entrained in the feed water. Death can normally be expected for any plankton that pass through the screens and enter the plant. The magnitude of these lethal effects is dependent on several factors, including the volume of water being withdrawn, the source of the water, the depth from which is pumped, and therefore the abundance of different types of marine life and the size or swimming ability of the organism.

In recent environmental impact assessments, impingement and entrainment issues have usually been considered the largest single ecological barriers to desalination plant siting. An important perspective should be kept in mind in this type of assessment, however, which is the volume of seawater used for cooling in existing coastal power plants. Most coastal states have sited their electrical generating stations along the coastline so they can use the very large volumes of easily accessible ocean water in the cooling process.

California, for example, has a number of large coastal power plants, and one of the largest is at Moss Landing along the Monterey Bay coastline of Central California (see chapter 9, figure 9.8). This plant, with a current capacity of 2,484 MW, has been operating for sixty-five years. Cooling water, which is withdrawn at a rate of about 1.2 bgd, is taken from either the adjacent Elkhorn Slough or Monterey Bay. Several desalination facilities have been proposed around the margins of Monterey Bay, one of which would pump about 5 mgd a day from the bay in order to produce about 2 mgd of freshwater. While the potential impingement or entrainment effects of this volume of water were raised as concerns, 5 mgd is only 0.4 percent of the volume of water that has been withdrawn by the power plant for decades. It is important to realize that power plant cooling water intakes usually operate at a much higher daily volume than any of the proposed desalination facilities, and there is much to be learned from the effects of these existing plants. Monterey Bay is a highly productive environment that still supports a healthy commercial and recreational fishery and is widely known for the diversity and abundance of its marine life.

Some field studies have recently been carried out around several large California coastal power plants to estimate the loss of plankton that takes place in the entrainment process. The constituents of the plankton include eggs and/or larvae of a number of marine invertebrates and fish, as well as the phytoplankton (diatoms and other algae, for example), and zooplankton (krill and other similar organisms), which vary in abundance depending on time of year, location, and water depth. Most of the plankton is concentrated in the upper 150 feet of the ocean, or the photic zone. The results of a field study at the existing Moss Landing power plant mentioned above (with a daily intake of 1.2 billion gallons) indicated that there was an estimated 13 to 28 percent loss of larvae over an area of 390 to 480 acres of the adjacent ocean. For some areal perspective, Monterey Bay has a surface area of over 100,000 acres. The proposed desalination plant would operate at less than 1 percent of the intake volume of the existing power plant.

These small and localized impingement and entrainment effects, however, can be significantly reduced in several ways.

1. *Subsurface intake wells:* Wells or intake pipes can be buried on the seafloor or placed beneath the beach, which can greatly reduce the uptake or entrapment of marine life. Sand can act as an additional filter such that most marine life is not affected. While this approach was originally believed to be a better alternative than pumping directly from the ocean, the sand can limit pumping rates, these intakes can interfere with freshwater aquifers, and many of these wells or systems have suffered from clogging and collapse.

2. *Deeper water intakes:* The great majority of plankton is concentrated in the photic zone, so placing an intake below this zone will significantly reduce the amount of impingement and entrainment and also will withdraw water that is cleaner and more uniform in quality. This will also increase construction costs, as intakes will need to extend farther offshore to reach deeper water.

3. *Multiple intake ports:* By withdrawing seawater from a number of inlet pipes or multiple ports from a single pipe, the velocities and volumes at any individual intake will be significantly reduced and therefore less hazardous to larger marine animals.

After the feeder or intake water has passed through the reverse osmosis or distillation process, about one-half of the water is fresh and the other half contains approximately twice as much dissolved salt as the original seawater. Typical or average seawater contains about 3.5 percent by weight salt, so what is commonly referred to as *brine* now contains about 7 percent dissolved salts and is therefore of slightly higher density. The actual salt composition is not different from the original seawater; only the concentration has increased. Seawater throughout the world oceans will contain the same percentages of all the major ions; the six most important and their abundance are chloride (55%), sodium (30.6%), sulfate (7.7%), magnesium (3.7%), calcium (1.2%), and potassium (1.1%). These six ions make up 99.3 percent of all the dissolved salt in seawater and is what is left on your skin when you get out of the ocean and dry off.

In addition to containing twice the amount of dissolved salt than typical seawater, the discharge from a plant will also contain any chemicals used in the desalination process. These may include chlorine or other disinfectants that are used to keep organisms from fouling or growing on filters, membranes, or pipes; anti-scalants to prevent salt buildup, coagulants to help flocculate particles or cause them to aggregate, and various other rinses or chemicals to clean the membranes. Many of these after various neutralizing or other treatment processes will be discharged with the brine into the receiving water, although their concentrations will be greatly diluted.

Direct ocean disposal is the most common discharge method for coastal desalination plants, and the brine can be released directly from the plant, combined with cooling water from an adjacent power plant, or mixed with treated wastewater from a nearby sewage treatment plant. Domestic wastewater treated to a secondary or more advanced level generally will have a very low dissolved salt content, so mixing this effluent roughly 1:1 with brine from a desalination plant can produce a discharge that has essentially the same salinity as seawater. The assumption is that ocean discharge will be diluted with a much larger volume of ocean water, which is the same assumption for treated wastewater, thereby reducing any environmental impacts. Dilution will take place with any fluid discharge into the coastal ocean. However, if the brine from a desalination facility is released directly from the plant, since it is denser than seawater it will initially remain on the bottom. The local oceanographic conditions, water depth, ocean currents, wave action and seafloor turbulence, will all affect the rate of dilution and therefore the seafloor area affected by the brine. The effects of any of the other residual chemicals contained in the discharge will depend on their concentration, their possible toxicity, the rate of dilution, and the marine life present in the discharge area.

There are a number of ways that any potential impacts of brine discharge can be reduced or mitigated. As discussed above, combining the brine with treated wastewater to reduce the salinity to match that of the receiving water body can dilute the salinity of the discharge to an acceptable or harmless level. The effluent can be dispersed through a series of diffusers spread out over a considerable distance of seafloor, which is often done with any ocean water discharge, and that will dilute the brine more quickly and spread it out over a much larger area rather than a point discharge. Siting the outfall or discharge point where turbulence from waves and currents will more effectively dilute the brine can also reduce any potential impacts.

## WHERE DO WE GO FROM HERE?
## SOME FINAL THOUGHTS

Desalination of coastal ocean water or brackish water from an estuary or bay has been utilized to produce freshwater in otherwise water-scarce regions for decades. There are many studies that have been performed and considerable data are available regarding intakes and discharges, impingement and entrainment, and the magnitude of the impacts and the areas affected. Desalination technology, whether distillation or reverse osmosis, is proven; technologies continue to improve, costs are known and coming down, and as many newer plants are now operating this isn't a new or unknown process. With growing coastal populations, increasing per capita use, and climate change, there aren't a lot of cheap and easy options to choose from for providing additional freshwater. There are very real construction and energy costs, but there will be similar costs for providing large volumes of additional water from any new source. There are potential marine impacts—impingement, entrainment, and discharge—but relative to the volumes of water being withdrawn for coastal power plants and the very large volumes of coastal ocean water desalination plant intake and discharges are small, and these impacts can be reduced and mitigated. Ocean water will always be available along any coastline in the world, independent of any future climate changes. This should provide us with some assurance that we do not need to ever run out of supplies of potential fresh water. However, whether reverse osmosis or distillation is the means used, significant electrical energy is required, and as long as we are still highly dependent on fossil fuels for our energy, large-scale desalination will contribute to our climate change problems.

# CHAPTER 13

# Carbon Dioxide, Climate Change, and Ocean Acidification

## INTRODUCTION TO GREENHOUSE GASES

A natural blanket of greenhouse gases surrounds the Earth and has made it far more livable than it would be otherwise. The average surface temperature of the planet is a reasonably comfortable 60°F rather than the 0°F it would be without our unique atmosphere. Carbon dioxide, methane, nitrous oxide, and fluorinated gases are the four major greenhouse gases, and each one is present in a different concentration. Each gas also has a different lifetime in the atmosphere, and each originally had its own natural concentration, but these have now been significantly increased by human activities. Water vapor is also a greenhouse gas, but its concentration in the atmosphere is not affected by human activities.

There is widespread agreement among climate scientists that greenhouse gases from human activities are the major drivers of the climate changes we have experienced since about the middle of the twentieth century. Greenhouse gas emissions from anthropogenic sources in the United States increased by 6 percent from 1990 to 2013, with electricity generation the single largest source of emissions, followed by transportation. The United States has begun to reduce its emissions, however, and since 2008 emissions have decreased by 9 percent, with emissions per person also dropping slightly. The global greenhouse gas situation is different, and from 1990 to 2010 net emissions increased by

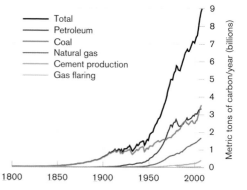

FIGURE 13.1. Changing contributions of different carbon sources to the atmosphere since 1800.

35 percent (figure 13.1). Carbon dioxide emissions, which account for about 83 percent of all greenhouse gases, increased 3.4-fold during this same period.

China is now the largest emitter of carbon dioxide, producing 26 percent of the total emissions, followed by the United States with 15 percent, the entire European Union with 10 percent, and India and the Russian Federation with about 6 percent each. These five nations or regions generate nearly two-thirds of the global carbon dioxide produced by human activities. China, however, through efforts to reduce air pollution and declines in manufacturing, has leveled off its carbon emissions, which had been growing at 8 percent per year, to a growth rate of almost zero. In early 2016 the central government in China told twenty-eight of its thirty-one mainland provinces not to approve any more coal-fired power plants. In addition, China's solar and wind energy capacity grew in 2015 at 74 percent and 34 percent, respectively, compared to 2014. All this is good news for the world's most populous country.

Carbon dioxide is the most abundant of the four greenhouse gases, making up about 76 percent of the total (figure 13.2). It enters the atmosphere from the burning of fossils fuels (oil, natural gas, and coal), as well as trees, wood products, and solid wastes, and also from certain industrial processes, a big one being the production of cement (figure 13.3). The average lifetime of $CO_2$ in the atmosphere is often listed as hundreds of years, but this is rather poorly defined because the gas isn't destroyed or broken down but rather through circulation through the atmosphere-ocean-terrestrial environments may be absorbed quickly or may remain in the atmosphere for thousands of years. In terms of relative potency over a hundred-year period, $CO_2$ is given an arbitrary value

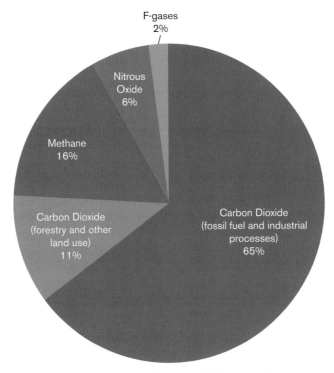

FIGURE 13.2. Percentage contributions of different greenhouse gases to the atmosphere (2010).

of one, and the other gases are usually compared numerically to carbon dioxide.

Methane ($CH_4$) enters the atmosphere from releases during the extraction of coal, natural gas, and oil but also from livestock and other agricultural activities, as well as the decay of organic material in landfills and from the thawing of permafrost. Methane is the second most abundant greenhouse gas, making up about 16 percent. It has an average lifetime in the atmosphere of 12 years but is about 28 times more potent as a greenhouse gas than $CO_2$.

Nitrous oxide ($N_2O$) constitutes about 6 percent of all greenhouse gases and is emitted during the burning of fossils fuels and organic wastes and also from various industrial and agricultural practices. Its atmospheric lifetime is about 121 years, and it is 265 times more potent than $CO_2$ in its heat-trapping role.

Fluorinated gases such as hydroflourocarbons, perfluorocarbons, and a few others are synthetic gases released during a number of different

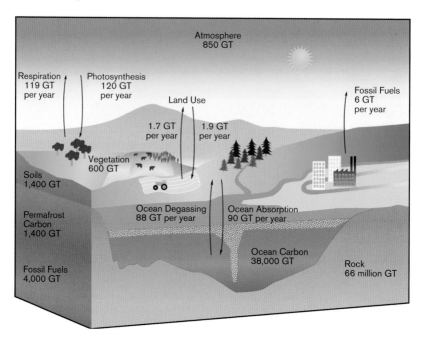

FIGURE 13.3. Sources and sinks for carbon and their average annual estimated volumes in gigatons per year.

industrial activities. They are potent or powerful greenhouse gases and make up about 2 percent of the total, and their average lifetime in the atmosphere varies from a few weeks to thousands of years. Their potency also ranges widely, with sulfur hexafluoride being the highest at approximately 23,500 times greater than carbon dioxide.

## INCREASING GREENHOUSE GASES IN THE ATMOSPHERE

Trying to determine just exactly how much greenhouse gas humanity emits every year, how much of each gas is produced by different sources, and then how much ends up in each different sink (ocean, atmosphere, and land/vegetation) is a nearly impossible task (see figure 13.3). Yet a lot of scientists have worked hard to try to estimate these amounts. As a result, there are a lot of big numbers out there—carbon dioxide released in gigatons per year, in metric tonnes per year, in pentagrams of carbon per year—and not surprisingly they don't all agree. I always have difficulty trying to envision what a ton or a gigaton of gas looks like. A ton of coal I can picture, but tons or pentagrams of carbon doesn't work in my head. The important concepts here are that we continue to release increasing greater amounts of $CO_2$ and other greenhouse gases each year, no matter what units we measure them in, and that there are very significant negative impacts from the accumulation of greenhouse gases in both the atmosphere and the oceans.

There is some general agreement that between 25 and 30 percent of all the $CO_2$ civilization has released over the past two hundred years, about 525 billion tons give or take a few billion tons, has been absorbed by the ocean. Because our annual global $CO_2$ emissions continue to increase, the oceans are taking up more and more $CO_2$ each year. The oceans are now absorbing about 1.3 million tons of carbon dioxide every hour. Fortunately for us, the oceans form a huge buffering system, which has reduced the amount of $CO_2$ that has gone into the atmosphere. Overall the oceans have about fifty times more $CO_2$ than the atmosphere, so they have a large capacity to both absorb and release carbon dioxide. There is an important balance here to keep in mind, however. The more carbon dioxide the oceans absorb, the less goes into the atmosphere. As a result, we slightly reduce the global warming and climate change impacts that additional $CO_2$ would have produced. On the other hand, the more $CO_2$ the oceans absorb, as discussed below, the greater the negative impacts on the ocean.

With a carbon dioxide content about fifty times greater than the atmosphere, even a modest release of the gas from the oceans could have a huge impact on the atmosphere and global temperatures. About 55.5 million years ago, there was just such a release, now known as the Paleocene-Eocene Thermal Maximum, or PETM, named after the geologic epochs when this globally devastating event occurred. The evidence for this upheaval in the ocean was first recovered in the deep-sea sediment record during an ocean-drilling expedition (figure 13.4). It is believed that a massive and rapid release of carbon from the ocean, in the form of both methane and carbon dioxide, entered the atmosphere, approximately doubling the $CO_2$ concentration, which caused global temperatures to rise by over 9°F. This was a period in Earth history when the Earth was already gradually warming, but this event, which still isn't well understood, pushed temperatures up even faster. The trigger may have been a massive release of *methane hydrates* (an icelike material found on the deep-sea floor) from seafloor sediments and also the release of greenhouse gases from thawing permafrost. This may have set in motion a set of positive feedbacks, where additional warming led to

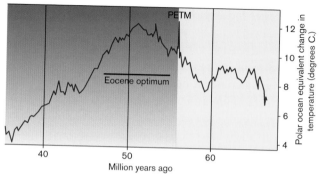

FIGURE 13.4. Global temperature increased significantly 55.5 million years ago (the Paleocene-Eocene Thermal Maximum) due to a massive carbon release from the ocean.

more greenhouse gas release from the ocean and more permafrost thawing, creating additional warming. Mass extinctions followed, and this very warm interval may have lasted over 150,000 years.

This completely natural event gives us some clear indications about what has happened in the geologic past when very large atmospheric and oceanic increases in carbon dioxide levels occurred. The total amount of carbon dioxide released during this event 55.5 million years ago is roughly equivalent to the amount that would be released by our burning all of the planet's remaining coal, oil, and natural gas, which we are proceeding to do at a rapid rate.

## INCREASING CARBON DIOXIDE IN SEAWATER AND OCEAN ACIDIFICATION

Ocean acidification (OA), which is beginning to have some significant effects on marine life and has the potential to have some very large future impacts, is a result of a fairly straightforward series of chemical reactions that occur when carbon dioxide enters the ocean. The initial step is the carbon dioxide reacting with seawater to form carbonic acid: $CO_2 + H_2O \rightarrow H_2CO_3$. The carbonic acid then rapidly disassociates to a bicarbonate ion and a free hydrogen ion: $H_2CO_3 \rightarrow HCO_3^- + H^+$. Some of the bicarbonate can react with hydrogen to form a carbonate ion and another hydrogen ion: $HCO_3^- + H^+ \rightarrow CO_3^{2-} + 2H^+$. These reactions can go in either direction, however, depending on the pH or abundance of $H^+$ ions.

The free hydrogen is what produces the H in pH. It is a measure of the acidity or alkalinity of a fluid. The more free hydrogen, the more acidic the seawater. To be clear, we are not talking about a strong acid like vinegar or lemon juice. In fact, the ocean is slightly alkaline, with an average pH of a little over 8.0, but it can vary a bit depending on depth and local ocean conditions. Average surface ocean pH has dropped from a "natural" preindustrial value of 8.2 to 8.1. Because the pH is measured on a logarithmic scale, this decline in the ocean pH of 0.1 unit corresponds to an overall increase in ocean acidity of 30 percent.

The last and very important reaction is the one that a wide variety of marine organisms use to produce their shells: $Ca^{+2} + CO_3^{2-} \rightarrow CaCO_3$. Organisms combine a calcium ion in seawater with a carbonate ion to form calcium carbonate. However, with much of the carbonate ion ($CO_3^{2-}$) often tied up in bicarbonate ($HCO_3^-$), it isn't always available for the formation of $CaCO_3$, which many marine organisms require for their shells or skeletons.

As carbon dioxide emissions continue with the burning of additional fossil fuels, more carbon dioxide will enter the oceans and pH will continue to decrease. Predictions at present are that pH is expected to fall to 7.8–7.9 by 2100, which would be equivalent to a doubling of acidity. Future rates of change will depend on when and how rapidly the United States and the rest of the industrialized world choose to move away from a fossil fuel–based economy, whether in the production of electricity, the types of fuels we use for transportation, or other uses. So far we are only taking very small steps in this direction.

## OCEAN ACIDIFICATION EFFECTS ON MARINE LIFE

The progressive decline in ocean pH, or shift toward increased acidity, will gradually begin to affect an increasingly diverse group of marine organisms. Interest and concern were initially focused on those organisms that build their skeletons or shells out of calcium carbonate. These include some of the larger and more visible animals such as coral, urchins, oysters, clams, and mussels but also many of the plankton, such as *foraminifera*, *coccolithophores*, and *pteropods*. Pteropods, also known as sea butterflies, are very small zooplankton that form thin calcium carbonate shells and are believed to make up about 45 percent of the diet of Pacific salmon, as well as providing food for a variety of seabirds and whales. Experiments done with these little creatures show direct correlation between seawater pH and shell formation or loss. While we can't see these little sea butterflies, and perhaps 99.9 percent of the people on Earth have never heard of them, a significant reduction in their population at the base of the food chain can have a very substantial effect on animals we can see and have heard about, salmon and seabirds, for example.

With continued research and monitoring, however, we have begun to develop a better understanding of how changes in seawater pH can affect a wide range of plants and animals in the ocean, not only those that build shells of $CaCO_3$. Calcium carbonate dissolves in acidic solutions, so the lower the pH, the more difficult it will be for any organisms that form shells out of calcium carbonate to either grow new shells or skeletons or maintain their existing health and populations. As is the case with most organisms, the larvae or the young in a population are the most sensitive to environmental changes. So whereas a mature, thick-shelled clam or oyster may be able to keep up with the dissolution brought on by a lower pH, the shellfish larvae, which are trying to grow a shell for the first time, may be simply out of luck. The biological impacts of ocean acidification will vary among different kinds of marine life as well, because they have a range of sensitivities to changing seawater chemistry. A tiny pteropod, which has a very thin shell compared to a clam or mussel, for example, is much more susceptible to increasing acidity than a larger, thick-shelled mollusk.

Beginning in about 2006, oyster larvae began undergoing dramatic die-offs in several commercial shellfish hatcheries in bays and estuaries along the Pacific Northwest coast of the United States. While some of the hatcheries had suffered losses of young oysters in the past, usually from some bacteria or other pathogen, those in 2006 were major mortalities. Seventy to 80 percent of the larvae had died and were accumulating at the bottoms of the growing tanks. With a larval die-off from a pathogen, disinfecting the tanks and filtering out the bacteria would usually solve the problem. But from 2006 to 2008 the disinfection and filtering did not halt the mortality events.

Two of these shellfish hatcheries alone produce the great majority of the oyster seed, or earliest life stage, on the West Coast. The oyster industry here has an estimated total economic impact of over $200 million annually, which makes up 80 percent of the shellfish produced in the area, generates 60 percent of the income, and provides over 3,000 jobs. These family businesses also raise clams and mussels and account for most of the West Coast commercial shellfish production.

After spawning, the larvae are grown for several weeks, and while still microscopic in size they are shipped to oyster farms in the United States, Canada, and Mexico, where they grow to maturity and are then marketed. But in 2007 many of the oyster larvae died before they could be shipped and the hatcheries couldn't fill their orders. With the help of a biogeochemist and marine ecologist, the oyster farmers discovered that the dissolved $CO_2$ in the seawater they were pumping into the hatchery was way off the normal pH scale. These mortality events, now nearly a decade ago, were perhaps the first clear indication that ocean acidification was not a distant future problem but was already having a serious impact on the Pacific Northwest's shellfish industry.

You might wonder why this would happen in the seemingly pristine waters off Oregon and Washington. Some scientific detective work soon began to put the pieces together. Along the West Coast wind patterns shift during late spring and early summer. Winds tend to blow from the northwest during these months and move surface water southward along the coast. When combined with the effects of the Earth's rotation, nearshore surface waters are transported offshore. Subsurface water migrates to the surface to replace the water moved offshore, in a process known as *upwelling*. And it turns out that the upwelled water off Oregon and Washington may actually have absorbed its dosage of carbon dioxide from fossil fuel combustion several decades earlier when it was at the sea surface. And it is this upwelled surface water that can hold more dissolved $CO_2$, which was being piped into the hatchery tanks. The reactions of the $CO_2$ with seawater begins the process of ocean acidification and reduces the availability of calcium carbonate that these tiny oyster larvae need to start creating their shells. The first few days of an oyster larva's life are critical as it starts to form its shell, and this is where nearly all of its energy goes. With less calcium carbonate available, it needs to use a lot more energy trying to build its shell. This leaves little or nothing to carry out its normal feeding, growth, and other physiological processes. This caused the massive die-offs, threatening the entire oyster industries of Oregon and Washington.

By monitoring water chemistry and with some scientific input, the growers learned that early in the morning the water was slightly more acidic, but by afternoon, when the phytoplankton had been actively photosynthesizing and utilizing $CO_2$, the $CO_2$ content was lower and the water less acidic (figure 13.8). They also realized that by adjusting the pH with sodium carbonate, they could help the tiny oysters in their earliest life stage. Adding to the problem in the bays where the shellfish were being grown was the occasional accumulation of algae and other plants, whose growth had been stimulated by an excess of nutrients, often from sewage or fertilizer runoff. The accumulation of dead algae becomes food for bacteria, which use up more oxygen breaking down the algae, increasing the carbon dioxide content further.

One of the Washington State family-owned oyster businesses borrowed money to move half of its production to Hilo on the Big Island of Hawai'i in order to escape the sometimes-lethal water conditions in Willapa Bay. The shellfish industry is also trying to develop more dissolution-resistant types of oysters.

Farther north, a team of NOAA scientists are concerned about the potential effects of OA on the $100 million annual Alaska king crab fishery. This crustacean, which may grow as large as six feet across, is a highly valued and expensive delicacy. As a way of looking into their future, when scientists exposed baby king crabs to $CO_2$ levels that are expected by midcentury, they died at twice the rate of those in normal seawater. Increasing the $CO_2$ to levels projected a few decades later led to young crabs dying in far greater numbers. While the experimental conditions were originally thought to be some time off in the future, conditions near the bottom of the Bering Sea, a primary habitat for king crabs, are now actually worse than that at certain times of the year. Although the lower pH may have affected shell construction, the effects seen in these experiments probably had more to do with the physiology of the young crabs and maintaining an appropriate acid-base balance, which becomes more difficult with ocean acidification.

Very similar research was carried out recently in a NOAA laboratory in Puget Sound with Dungeness crabs. Crab eggs were collected and placed in tanks of seawater at different pH levels. While larvae hatched out at the same rates in all tanks, the survival rate for those in water of pH 8.0, roughly what occurs today, was 58 percent for the first forty-five days. At pH 7.5, which sometimes occurs in Puget Sound today, the survival rate dropped to 14 percent. The Dungeness crab fishery in 2014 was valued at a combined $195 million for California, Oregon, and Washington, so there are very substantial concerns for the long-term survival of this crustacean and the industry.

The issue of ocean acidification goes well beyond the king and Dungeness crab fisheries and the shellfish hatcheries of the Pacific Northwest. A 2007 oceanographic voyage along the Pacific Coast from British Columbia to Baja California discovered that corrosive waters were widespread in this upwelling region where colder, low pH water is seasonally brought to the surface. Similar conditions have also shown up along the Atlantic coast of the United States. Sewage discharge, agricultural runoff, and rising acidity have begun affecting the formerly thriving oyster industry in Chesapeake Bay as well.

The impacts on the economically important oyster industry in the Pacific Northwest have triggered more investigations and have brought attention to other, subtler effects of ocean acidification. A more acidic ocean alters ocean acoustics, with sound absorption decreasing 40 percent with a pH decline of 0.3, which has been predicted by 2100. As a result, sound will propagate farther and will add to an already noisy ocean, which will most likely affect marine mammals that are highly dependent on acoustics.

The market squid are a keystone, or central, species in the ocean in that they either eat or are eaten by just about everything. These cuttlefish make up California's biggest fishery, about two-thirds of the entire tonnage of the commercial catch year after year. Anything that happens to the squid population will affect the entire food chain. Scientists at the Woods Hole Oceanographic Institution carried out experiments to see how a more acidic ocean would affect the squid. After fertilized egg capsules were released by mating squid, one batch was kept in normal seawater and a second was transferred to a tank with a pH about three times more acidic, conditions again projected to occur by about 2100. The squid in the lower pH water took longer to develop and were about 5 percent smaller than the others. In addition, their statoliths (the equivalent to human inner ear bones that control our balance) were smaller and abnormal, which disrupted their balance and caused them to swim in circles. An increase in OA has also been shown to reduce metabolic rates in jumbo squid, depress the immune response of blue mussels, and affect the sense of smell in some fish.

While seagrass and algae can benefit from ocean acidification, most of the research shows that the benefits they derive are offset by the negative impacts on the species living within seagrass habitats. There may well be far more negative impacts of ocean acidification that we simply have not observed or investigated yet.

The synergistic effects of both a warmer and a more acidic ocean may also be gradually beginning to affect other marine organisms. Coral growth on the world's largest coral reef, the Great Barrier Reef off northeastern Australia, has experienced a decline in its growth rate over the past twenty-five years, and the coral also appear to be thinner and more brittle. When more energy is spent maintaining a skeleton or a shell, the organism has less energy to expend on other important processes such as growing, resisting disease, reproducing, or responding to stresses. Nearly 30 percent of the ocean's tropical corals have disappeared since 1980, mainly due to more frequent and intense warming events, but an increasingly acidic ocean is likely to compound these effects. Coral reefs are the basic foundation for the lives of over 100 million people in tropical latitudes and have multiple benefits including fisheries, recreation and tourism, and coastal protection. This is discussed in chapter 15.

WHERE DO WE GO FROM HERE?

There are numerous negative biological impacts of ocean acidification that have already been recognized, and also many laboratory experiments that have and are being carried out to help us understand how different marine organisms will respond to projected future pH conditions. While these low pH values may still be some decades away, the warning light is already on. Marine species that inhabit coastal waters have, in many cases, evolved to tolerate rapid environmental changes, intertidal organisms, for example, but the current rate of acidification is unprecedented and estimated to be ten to one hundred times more rapid than at any time in the past 50 million years.

The continuing increase in acidification in coastal waters in the decades ahead has enormous potential implications for ecosystems, in particular, those of commercial and economic importance, but also for organisms at the base of the oceanic food chain. Evidence gathered to date indicates future impacts will be substantial, although it is difficult to predict with a high degree of confidence when we will reach a tipping point with any particular organism. Unfortunately, while we can easily measure the pH of the ocean and we know it has increased 30 percent, we can't see it. An analogy is taking the rivets out of an airplane flying at 35,000 feet, one by one; the first few or even perhaps the first dozen aren't noticed and may not have a significant impact, but at some point, we reach the point of no return and the game is over. We don't ever want to get to this point, and we need to start significantly reducing the entire planet's dependence on and consumption of fossil fuels now, for reasons beyond ocean acidification.

Recovery of the ocean's pH to preindustrial levels is likely going to take decades to perhaps centuries. The fundamental and well-understood cause of ocean acidification is the emission of increasing amounts of carbon dioxide, with about 25 to 30 percent of this, 1.3 million tons per hour, ultimately absorbed by the oceans. Anything we do to turn this increasing ocean acidification around is also going to take many decades or longer, even if we start now. There are some additional steps that can help some coastal regions: controlling or reducing nutrient runoff from sewage disposal as well as the application and then runoff of excess agricultural fertilizers. Carbon dioxide is also captured by plants and stored, and in coastal environments this includes mangroves, salt marshes, and seagrass beds. These ecosystems need to be protected, preserved, and expanded where possible.

But the 800-pound gorilla in the room is our burning about a million tons of coal, nearly 4 million barrels of oil, and 15 billion cubic feet of natural gas, every hour, all day long, 365 days a year. This is not sustainable. And it is having widespread impacts on the coastal oceans that need to be recognized and mitigated now. An encouraging sign was the global investment of about $289 billion in renewable energy in 2015, combined with more new renewable energy capacity than conventional capacity added for the first time in history.

CHAPTER 14

# Coral Reefs and Threats to their Health and Survival

Corals first appeared in the evolution of the Earth over 500 million years ago in the Cambrian period, but it took another 100 million years or so until they became widespread and common in the fossil record. After millions of years of building and surviving mass extinctions and glacial and interglacial periods with rising and falling sea levels, these simple organisms and the very large and complex ecosystems and structures they have created seem to be facing the most serious threats from human activities across the planet.

Australia's Great Barrier Reef, about 1,200 miles long and 90 miles wide, is usually recognized as the largest structure on the planet built completely by living organisms. Although coral polyps are tiny and lack backbones, they build rock-hard structures of calcium carbonate that can survive storm waves, typhoons, tsunamis, and a number of other natural processes, but they are not immune to the activities of humans.

There are two quite different types of corals:

- *Hermatypic.* These are the reef-building colonial corals and are unique in having a symbiotic relationship with algae (dinoflagellates), which actually grow within the coral polyp. This is usually a very productive relationship, with the algae using carbon dioxide, water, and a few key nutrients to produce food and oxygen through the process of photosynthesis. The other half of this partnership, the coral (an animal), extracts microscopic plankton from seawater, utilizes the oxygen and the food produced by the algae, and through respiration produces carbon dioxide. This cozy relationship has worked quite well for hundreds of millions of years as evidenced by preserved coral reefs throughout the fossil record.

- *Ahermatypic.* These are solitary corals that have no symbiotic algae and therefore are not restricted to the near-surface waters where light can penetrate. Solitary corals can live in very deep cold water in utter darkness, in fact as deep as 20,000 feet, and form what have relatively recently been recognized as lush coral gardens, with nearly as much species diversity as reef-forming corals. Having no symbiotic algae, these corals gain all of their nutrition by capturing small plankton with their tentacles.

The hermatypic, or reef-building, corals have a few very important restrictions on where they can grow and therefore where reefs exist today. They generally do best in warm, sunny, clear, and shallow water, as follows:

- *Temperature:* Year-round ocean temperatures between about 68°F and 86°F (20°–30°C) are optimal, which generally restricts healthy coral growth to the regions of the ocean between about 30° north and south of the equator.
- *Sunlight and Depth:* Because hermatypic corals have symbiotic algae, they need to be within the photic zone, or the upper approximately 150 feet of water, where enough sunlight penetrates to allow for photosynthesis.
- *Salinity:* Corals are most productive at near-normal ocean salinities and can do well in salinity between about 32 and 37 parts per thousand, or 3.2 to 3.7 percent salt. Reefs are generally lacking directly offshore from where rivers enter the ocean and create a persistent freshwater plume.
- *Turbidity and Suspended Sediment:* Excessive suspended sediment from rivers, or persistent high turbidity, perhaps from plankton, will diminish and restrict coral growth, either by limiting light penetration or by literally smothering the coral polyps in sediment.

In addition to these critical requirements, corals need a hard substrate to settle onto during their larval stage in order to transition to a calcium carbonate secreting stage. An existing reef, crustose coralline

FIGURE 14.1. A luxuriant and healthy coral reef ecosystem in Indonesia. (Photo: Tom Gruber © 2012)

algae, another coral, or other hard-shelled organism works fine. Because the symbiotic algae living within the coral need nutrients and carbon dioxide and the coral itself needs oxygen, reefs grow best in areas of active water circulation, whether wave or current activity or both.

## IMPORTANCE AND VALUE OF CORAL REEFS

Reefs are now recognized by most people as having many benefits and playing important geologic, biological, and economic roles throughout the tropical oceans of the world. Most marine biologists consider coral reefs to be the most diverse communities in the oceans, in part because they foster and host a huge diversity of species, surpassing even that of the tropical rain forests. Reefs provide shelter, food, and breeding and nursery grounds for an estimated 35,000 to 60,000 species worldwide, including about one-third of the world's estimated 25,000 to 30,000 species of marine fishes (figure 14.1).

The economic value of reefs has become more widely appreciated in recent decades, especially as their health in many areas has been compromised. Coral reef fisheries provide revenue for many local communities as well as a global fishing fleet. Perhaps a billion people around the world depend to some degree on coral reefs for food and for fishing income, and it is believed that if properly managed and not overfished these environments could provide about six tons of fish per square mile each year. But that turns out to be a very big if.

Visitors to tropical areas and income from tourism have grown substantially in recent decades, and it is estimated that many countries, island nations, or areas with reefs derive over 50 percent of their overall income from tourism. Tourism is the largest component of the economy of Hawai'i, providing 21 percent of the income—about $14 billion in 2014—and over 150,000 jobs. Australia's Commonwealth Scientific and Industrial Research Organization (CSIRO) reported that nearly $100 billion in tourism was generated annually from the Great Barrier Reef, Florida reefs, and Caribbean reefs in the 1990s. In many small island nations where there is not a significant industrial base, coral reef–based tourism can significantly increase local sources of income, but this tourism depends on healthy and vibrant reefs and clear, clean water.

Coral reefs also can serve an important role in providing protection for coastal communities, development, and infrastructure by resisting the attack of storm waves, hurricanes, and typhoons. The buffer or barrier provided by strong, healthy reefs helps reduce shoreline erosion and coastal flooding and comes at no cost. Reefs can also adjust to rising sea levels because corals are generally able to grow upward, keeping pace with sea-level rise.

The increasing number of bioactive compounds of potential medical value derived from tropical ocean organisms has led to the development of a new field, marine pharmacology. While it is a very long road from the extraction of an organic compound from a sponge that grows on a reef to direct application in a doctor's office or hospital, a number of compounds have shown promise and are being used in treatments for diseases like cancer and HIV, among others.

In addition to their value in economic terms, coral reefs are the basis of the history, culture, and daily lives of societies that for centuries have inhabited tropical latitudes. Further, coral reefs have immense value in their own right, although we have been slow to recognize this and to understand the negative impacts of our expanding populations and their demands on the planet. The area of maximum coral reef biodiversity has been designated the Coral Triangle, a roughly triangular area that includes the tropical waters of Indonesia, Malaysia, Papua New Guinea, the Philippines, the Solomon Islands, and Timor-Leste. This is an area surrounded by many third world nations where conservation is economically challenging.

THREATS TO CORAL REEFS

*Water Pollution, Water Quality Decline, and Marine Debris*

Pollution has been recognized for some time as one of the leading causes of coral reef degradation, and these inputs or threats can come from a number of different potential sources (figure 14.2). Population growth and an expanding tropical tourism industry means more wastewater delivered to the coastal waters, whether pumped directly into the ocean or reaching the ocean through a river, estuary, or other terrestrial water source. Whether treated to some level or untreated, domestic wastewater and agricultural runoff, with its fertilizer content, will increase the level of nitrogen around offshore reefs, often causing an overgrowth of algae, which can literally smother reefs by cutting off their sunlight. Agricultural runoff may also include pesticides, which can take a toll on often-delicate reef organisms. Plastic, Styrofoam, and other marine debris is a problem that is becoming more widespread but also more publicly noticed (see chapter 9). Floating trash can cover reefs, blocking off sunlight that polyps need to survive. Turtles often mistake plastic bags for jellyfish and eat them, blocking their digestive tract and causing starvation. Seabirds, such as the albatrosses, are attracted by brightly colored plastic, whether floating or along the shoreline, and when consumed in large quantities, often leads to the death of these large and magnificent birds. Lost or discarded fishing nets—called "ghost nets"—made of non-biodegradable nylon, can snag on reefs and strangle thousands of fish, sea turtles, and marine mammals, for decades.

*Overfishing and Destructive Fishing Practices*

As stocks of fish have declined in tropical areas and new markets have opened up, the fishing methods utilized have become more extreme and damaging to the reefs. It is believed that about 50 percent of the world's coral reefs are threatened today by some combination of destructive fishing practices. Some of the current practices used on reefs are simply unsustainable, and the list includes fishing with poisons, blast fishing, and overfishing. Sodium cyanide is a broad-spectrum poison, which has been used in fumigating, in the electroplating and mining industries, and also for executing criminals. Its misuse in fisheries began in the early 1960s when the demand for fish for U.S. aquariums grew, making them the leading importer of coral and reef fish. The other demand that accelerated was the live fish market in restaurants in Hong Kong,

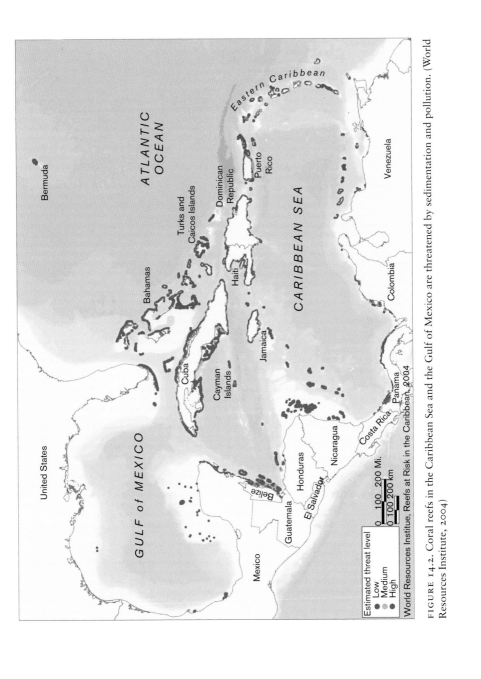

FIGURE 14.2. Coral reefs in the Caribbean Sea and the Gulf of Mexico are threatened by sedimentation and pollution. (World Resources Institute, 2004)

FIGURE 14.3. Live fish are kept in crowded tanks in some restaurants in Asia and command a very high price from diners. (Photo: Gary Griggs © 2011)

Taiwan, and other places in Asia with large Chinese populations. Fish are selected and then plucked live from restaurant aquariums, with some species bringing up to $300 a plate (figure 14.3). This is popular as a status symbol for business events and quickly became a billion-dollar-a-year business.

Divers, mostly from the Philippines, squirt cyanide into crevices in the reef, which stun the fish, making them easier to catch. Although some of the larger tropical fish can metabolize cyanide, the chemical cloud produced during this process poisons smaller fish and other marine animals such as coral polyps.

Despite the general outlawing of this practice throughout most Indo-Pacific nations, because of the high prices brought by the fish, weak enforcement capabilities, and corruption, the use has spread in reefs throughout the Philippines and Indonesia. As fish populations decline in one area, the illegal fishing moves on to a new area. The Philippines, home to most of the cyanide fishermen, has taken concrete action against this destructive activity by partnering with nongovernmental organizations and training fishermen to utilize alternative fishing methods, fine-mesh nets, for example. Indonesia has taken similar steps. More aggressive enforcement has also been implemented whereby exported fish are randomly analyzed for cyanide. In addition, there are new regulations that are tightening controls on the importation and distribution of cyanide, and there are public awareness campaigns to educate the people of the Philippines on the destruction created by this industry and its long-term impact on their reefs.

Other damaging fishing practices, such as banging on the reef with sticks, can destroy coral formations that normally serve as fish habitat. In some instances, explosives are used (blast fishing), which kill the fish by shock waves from the blast, leaving the fish to be easily picked up from the bottom or the ocean surface. Not only are the fish themselves killed, but the blasts kill other marine life in the blast area and also can destroy or damage living portions of the reef. While illegal, dynamite or explosive fishing still takes place in as many as thirty nations in the Indo-Pacific region. Once damaged, the reef will not recover for many years.

Overfishing is another leading cause of coral reef degradation. Many coastal communities in the tropical regions of the world depend on reef fisheries for their food supply as well as their livelihood. In many cases, these fisheries have been locally managed and maintained at sustainable levels for centuries because the local people understood the concept of sustainability and the need to leave enough fish for successful reproduction and growth in future years. However, many fish stocks are now being overfished and threatened by a combination of increasing demand for fish, lack of alternative employment options, nonlocal fishing boats or fishermen, more efficient fishing methods, and inadequate management and enforcement.

An additional example of a destructive practice is overfishing of herbivores, such as parrotfishes. Where reefs have been damaged or bleached, herbivorous fish, if present, can control the growth of algae, which can cover the reef surface and restrict coral regrowth. This has been the case in the Caribbean, whereas in Moorea parrotfishes are present and have kept the algae under control so reefs have been resilient.

In many cases, the fishermen doing the work, whether using cyanide or explosives or other destructive methods, make very little money and are exposed to significant risks of injuries and imprisonment while the middlemen who control the industry make the profits. It has been repeatedly shown that the tourism and recreational diving industry can flourish where reefs are alive and healthy. It can sustain itself year after year and can generate more stable employment and income for local

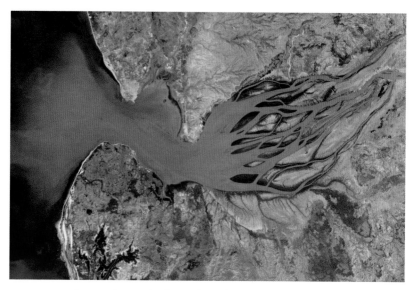

FIGURE 14.4. Large volumes of sediment from the Betsiboka River, Madagascar, enter Bombetoka Bay and then the Indian Ocean, where it may eventually settle onto coral reefs. (Photo: NASA)

populations than the destructive reef fishing practices that have proven so damaging to so many formerly pristine tropical areas.

### Sedimentation

Where rivers drain to coasts fringed by reefs and land use changes in the watersheds have exposed soil and led to increased erosion rates, silt and clay carried downstream may eventually end up on the reef (figure 14.4). Logging, vegetation clearing and burning, mining, and construction can all lead to soil loss and sediment deposition in coastal waters. While coral may be able to survive an occasional small amount of sediment, an excess may lead to death of the coral by smothering it and depriving it of both adequate light for photosynthesis, but also can significantly reduce the ability of the coral to feed from the overlying water. Mangrove trees and seagrass, which normally act as filters or traps for terrestrially derived sediment in some areas, are also being rapidly destroyed, which has increased the amount of sediment reaching some reefs. Mangrove forests are often cut for firewood or removed to create open beaches, and have also been removed so fish or prawn farms can be constructed. There are efforts underway in some tropical nations or states to restore these valuable filters, which also protect the shoreline from wave attack.

## Blasting, Dredging and Filling for Coastal Construction

As coastal populations have risen and tourism has grown, the demand on coastal resources in the tropics has increased as it has elsewhere around the world. A major difference in the tropics, however, is that the nearshore and, often, offshore areas are alive and typically consist of fringing reefs, barrier reefs, or atolls. Whether for harbors, shipping channels, airports, fill for housing, commercial or industrial development, construction has often led to direct loss of reefs. At one time in the past, big cities such as Hong Kong, Singapore, Manila, and Honolulu had thriving coral reefs. Long ago these reefs were destroyed by human construction and development. Now reefs growing near other coastal cities and communities are experiencing the same degradation and threats. Along many tropical islands and coasts, the only building material is coral, whether the active reef or the sand or mud on the seafloor that consists of broken-down reef material. As a result, the coral or coral sand is mined or extracted to use for roads, buildings, and cement.

Coral reefs are old and fragile and grow very slowly. They have been blasted, dredged, and buried around the world for development, often destroyed to build the airports or the harbors that allow airplanes to land or cruise ships to dock and deliver the people who come to see and experience the very reefs that are being damaged or destroyed.

Healthy reefs surrounded Singapore until the 1950s and 1960s when the government embarked on a national program of coastal land reclamation in order to expand its limited area. Over the past half century Singapore has expanded its total land area an amazing 22 percent by obtaining rock, earth, or sand from neighboring countries to extend the coastline outward and add to their existing land base (figure 14.5). This led to completely covering over fringing and patch reefs as foundations for major port facilities, petrochemical industries, military training grounds, hotels, and parks. The species of reef-forming corals had declined from thirty to eight by 1968. This continues today with sand being imported from other Southeast Asian countries in order to further expand their small territories by pushing the shoreline out.

Coastal land is at a premium, typically the most expensive property on the planet, and with a global economy now 95 percent dependent on marine transportation, giant container ports, oil refineries and terminals,

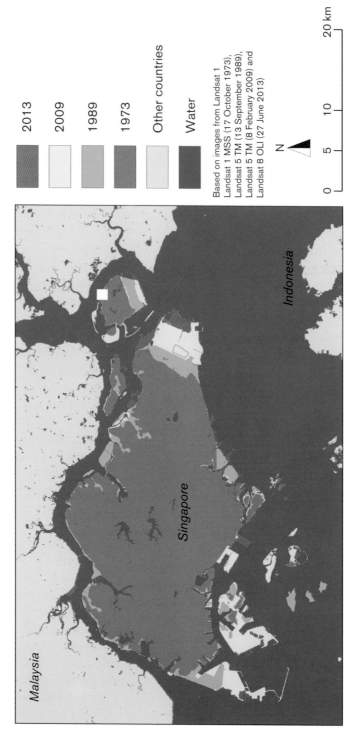

2013

2009

1989

1973

Other countries

Water

Based on images from Landsat 1
Landsat 1 MSS (17 October 1973),
Landsat 5 TM (13 September 1989),
Landsat 5 TM (8 February 2009) and
Landsat 8 OLI (27 June 2013)

N

0    5    10        20 km

Malaysia

Singapore

Indonesia

FIGURE 14.5. The land area of Singapore has been significantly expanded through artificial fill and reclamation. (Courtesy UNEP/GRID-Geneva)

cruise ship ports, and international airports are all expanding into tropical waters, with consequent loss of once-healthy reefs. In all likelihood, none of these facilities or infrastructure will be removed in our lifetimes, although sea-level rise in the long-term future may make the decision for us. One long-term approach is to utilize marine zoning to establish areas that are suitable for certain types of use or for protection or preservation rather than selling them for development to the highest bidder. The Earth's coastlines are replete with unfortunate examples where development has occurred with no planning or land use designations, or a complete lack of enforcement of any existing regulations or practices. These practices are often driven by a desire for more income, which can be realized by auctioning off or selling shoreline land. A combination of efforts to modernize and expand economic development opportunities, corrupt government officials, and a general lack of appreciation for the value of coral reefs and the many benefits they provide has led many nations or regions down a path to destruction of these important habitats and ecosystems.

## Tourism: Hotels, Boats, and Divers

Tourist resorts seem to be expanding exponentially in formerly pristine tropical areas, and many of these are virtually self-contained cities, with hundreds or thousands of people visiting and employed. While these clearly contribute very significant income and employment to the local economies, there is also the potential for substantial environmental impact. Although eco-resorts have appeared in recent years that are designed to significantly reduce environmental impacts, many of the huge existing hotel and resort complexes are quite a different story.

One of the most obvious potential impacts is the treatment and disposal of the sewage generated by all those people, whether discharged directly into the coastal waters or into septic systems, with seepage that may eventually reach the adjacent ocean. The effluent introduces nutrients, which fertilize algae and other plants that can literally grow over the reef corals. The treatment and release of wastewater must be very carefully engineered and planned if these negative impacts are to be avoided and the reefs preserved. The level of treatment as well as wave activity and circulation or the degree of natural flushing at the discharge point have an important influence on the impact of the effluent. High-level treatment along with onshore evaporation or percolation ponds or very long offshore outfalls are two possible solutions.

In addition to the direct effects of the countless tourists that stay in large resorts complexes in places like Hawai'i, Cancun, Tahiti, Fiji, the Bahamas, and Bermuda, many studies have documented the impacts to the offshore reefs from divers, snorkelers, dive boats, and cruise ships. Scuba divers potentially have far greater impacts on corals than snorkelers, who are usually swimming along at the surface. Direct physical damage to reefs includes accidental or intentional breaking of branched coral and trampling or walking on the reef surface, particularly at shoreline access points. Dive or fishing boats can damage reefs by breaking off coral when their anchors drag or are pulled out. Larger boats with bigger anchors can cause even greater damage. Areas of heavy recreational boating, diving, or fishing will obviously experience more serious consequences than areas with only occasional use. Installing permanent moorings where boats can tie up can help eliminate or reduce damage from anchors, and training and briefing of divers prior to water entry can also minimize damage from divers themselves.

## Ocean Warming and Coral Bleaching

Sustained warmer water temperatures can result in coral bleaching. Normal corals are greenish, yellowish, or gold to brown in color, due to the symbiotic algae living within the polyps. These algae, called zooxanthellae, normally provide the coral with up to 80 to 90 percent of their energy through photosynthesis, making them essential for coral survival and growth. The algae are also normally responsible for the color of coral. When corals experience stress, typically when the water is much warmer than normal, they can expel their algae, for unknown reasons, and turn white, or become "bleached" (they are, in fact, transparent, and one can then see the white skeleton through the tissue) (figure 14.6). There is a chance that bleached coral can survive and recover if conditions return to normal quickly enough. However, in the face of other human-induced pressures, corals have become vulnerable. In many cases, bleached coral colonies die. Algae can cover the dead coral, which makes recolonization by young coral larvae more difficult.

Global warming is increasing the surface temperature of the oceans as well as the atmosphere. On shorter time scales, El Niño events lead to changes in ocean circulation and water temperatures over large areas of the tropical oceans that may last for several months or longer. It has been these events in recent decades that have produced an increasing number of prolonged and widespread bleaching events.

FIGURE 14.6. Bleaching of staghorn coral at Palmyra Atoll National Wildlife Refuge in the Pacific. (Photo: Amanda Meyer/USFWS licensed under CC BY 2.0 via Flickr)

Coral bleaching is not new but has been recognized by scientists for nearly a century. Prior to the 1980s, when coral reef research intensified, localized bleaching incidents were reported and attributed to such factors as extremely low tides, hurricane damage, torrential rainstorms, freshwater runoff near reefs, or toxic algal blooms. Over the past thirty-five years or so, however, all major bleaching events have been associated with periods of exceptionally warm sea surface temperatures. These correlations between rising global temperatures and the occurrence of the two strongest El Niño events in the past century left little doubt that the unprecedented coral bleaching and mortality being observed today is linked to climate change.

Coral reef degradation has accelerated on a global scale since the late 1970s to early 1980s. In the 103 years between 1876 and 1979, there were only three bleaching events reported anywhere, although there were no doubt far fewer observers or scientists in the water during the earliest years. In the 11 years between 1979 and 1990, however, over sixty bleaching events were recorded. Many of these, 1982–83, 1987, 1997–98, 1995, and 2002, were El Niño years, with elevated water temperatures. In 1987 the first recorded major bleaching event occurred in the Caribbean. During the 1997–98 El Niño, massive bleaching occurred along the entire length of the largest barrier reef in the northern hemisphere, near Belize. One hundred percent mortality was observed in some areas with no recovery; instead sponges and algae grew over the reef. In Australia global mass bleaching was observed over 50 percent of the reefs during that event, and sea surface temperatures were the highest ever recorded. Assessments made in late 2000 showed that 27 percent of the world's reefs had been effectively lost, with the largest single cause being the massive climate-related coral bleaching event of 1998. This destroyed about 16 percent of the coral reefs of the world in nine months during the largest El Niño climate event ever recorded. Continued surveys in the ten following years, however, indicated that the coral reefs in the Indo-Pacific region had largely recovered (more quickly and completely than expected) from the devastating 1998 event that killed up to 90 percent of the corals on some reefs.

In 2005 the United States lost half of its coral reefs in the Caribbean in one year due to a massive bleaching event. The warm waters centered on the northern Antilles near the Virgin Islands and Puerto Rico and expanded southward. Comparison of satellite data from the previous twenty years confirmed that thermal stress from the 2005 event was greater than that of the previous twenty years combined.

The geographic extent of bleaching seems to expand with each new El Niño, particularly when combined with long-term global warming. Fifteen of the sixteen warmest years on Earth since record keeping began in the late 1800s have all occurred since 2000. Record hot temperatures in the Pacific Ocean in 2015 have fueled the worst coral-bleaching event ever witnessed along the northern third of Australia's Great Barrier Reef. From the air or underwater, gleaming pale white skeletons dominate the seascape. Aerial surveys have shown 95 percent of the Great Barrier Reef's northern reefs were rated as severely bleached, and only 5 of 520 reefs surveyed were unaffected by bleaching. The symbiotic relationship between the coral and algae is a somewhat fragile one, and algae don't like it hot. Unfortunately, all signs are pointing in the direction of warmer waters in the future. This is a huge problem for the tropical oceans of the world, and not one with a simple or local solution.

*Ocean Acidification*

Ocean acidification was discussed in some detail in the previous chapter, so with that introduction and foundation, the intent here is to focus specifically on what we know about the effects of ocean acidification on coral reefs. Since the beginning of the industrial revolution, the oceans have absorbed approximately 525 billion tons of carbon dioxide from the atmosphere, or about one-third of the anthropogenic carbon emissions released. While the uptake of carbon dioxide has significantly reduced the greenhouse gas levels in the atmosphere and minimized some of the impacts of global warming, the ocean's uptake of carbon dioxide is having negative impacts on the chemistry and biology of the oceans. The pH of ocean surface waters has already decreased by about 0.1 unit from an average of about 8.21 to 8.10 since the beginning of the industrial revolution. Estimates of future atmospheric and oceanic carbon dioxide concentrations suggest that by the middle of this century, atmospheric carbon dioxide levels could reach more than 500 ppm, and near the end of the century they could be over 800 ppm. This would result in an additional surface water pH decrease of approximately 0.3 pH units by 2100.

Corals build their skeletons out of calcium carbonate, just like shellfish and many different plankton. They need to have calcium and carbonate ions available in seawater in order to build reefs. This is particularly important in the earliest and most fragile life stages where a coral larva is ready to settle down on a hard surface and begin to metamorphose into a small polyp and grow a skeleton. Where lower pH levels are present, some studies have shown a 50 to 75 percent decline in larval settlement. Other studies have shown that mature corals calcify or grow their skeletons at reduced rates with lower pH waters, thereby making the corals more brittle and subject to other threats. While we refer to tropical reefs as "coral reefs," they are in fact composed of a large number of other calcium carbonate–secreting organisms, coralline algae, mollusks, sea urchins, clams, bryozoans, and foraminifera, among others, which are all affected by more acidic waters.

Coral reefs were, in fact, among the first marine ecosystems to be recognized as vulnerable to ocean acidification. While individual species vary in their response to changing pH, a number of different laboratory experiments show that a doubling of the carbon dioxide in the atmosphere (and we are now at about a 40 percent increase over preindustrial levels), which translates to a lower pH in seawater, would result in a 10 to 50 percent decline in the rate of calcification of reef-building corals. With more energy having to go into calcification, less energy is available for other physiological processes. Think about early humans having to invest almost all of their time and energy in hunting for food; there was little left over for anything else, and progress and change came very slowly.

With less calcification, there is less coral growth or accretion, and overall we end up with a less healthy reef. Over time, if this trend continues, coral growth decreases and the coral become weaker and will be more easily damaged or destroyed. Over time, coral itself will likely be outcompeted by other species, typically large algae. This often happens with damaged or weakened reefs today, where macroalgae grow over the coral skeletons and leave no surface exposed where coral larvae can settle to reestablish themselves. Without the living coral, a reef's biodiversity is diminished, and other reef organisms, like fish, no longer find this a viable habitat. In turn, the lack of coral reefs will also reduce the carbon dioxide buffering capacity of the ocean, thus increasing ocean acidification at an even faster rate.

This is a much bigger issue than simply an interesting and very large-scale biological experiment. Tourism and recreation in reef areas is based primarily on the coral reefs and the diverse and colorful fish that inhabit these environments. Without a healthy and vibrant reef, tourism and fisheries begin to decline and a significant portion of the local economy and food supply disappears. Strong, healthy growing reefs also are the best natural barriers or buffer to wave attack, storms, hurricanes or typhoons. Without them, communities and development in tropical regions will be more exposed and threatened.

Were the ocean's trend toward increasing acidification to be reversed, the reduction in calcification could be slowed and eventually reversed. Unfortunately, however, there is little hope of reducing the acidification anytime soon due to the enormity of the system—atmosphere and ocean—and the high carbon dioxide levels that are already embedded in both.

## WHERE DO WE GO FROM HERE?

Coral reefs historically have formed the fundamental biological, geologic, and economic framework for the coastlines of many countries and small island nations in the tropical latitudes around the world. Fisheries and tourism have been at the core of the economies of many of these areas, but development of many types, from dredging and blasting for harbors to filling for airports and industrial or tourist development, has led to damage and destruction of reefs over large areas. The direct and indirect impacts of destructive fishing practices, land use changes that affect coastal waters, tourist activities and their impacts, combined with global changes in ocean temperature and chemistry have all taken their toll on these fragile environments and important ecosystems.

What is the global status of reef health? The most recent (2008) State of Coral Reef Ecosystems Report developed by NOAA is an ongoing effort to evaluate the condition of the nation's shallow water coral reef ecosystems. The report contains studies of fifteen locations or jurisdictions within the United States and the Freely Associated States (FAS) in the Caribbean Sea and Pacific Ocean. Highlights of the report follow.

- Approximately half of the coral reef ecosystem resources under U.S. or FAS jurisdiction are considered to be in poor or fair condition and have worsened over time due to several natural and anthropogenic threats (see figure 14.2).
- Reef habitats adjacent to populated areas tend to experience more intense threats from issues like coastal development and recreational use, but even remote reefs far from human settlements are imperiled by illegal fishing, marine debris, and climate-related impacts such as bleaching, disease, and acidification.

The same NOAA report also evaluated the status of global coral reef ecosystems and includes additional highlights.

- The world has effectively lost 19 percent of the original area of coral reefs; 15 percent are seriously threatened with loss within the next 10 to 20 years, and 20 percent are under threat of loss within 20 to 40 years. The latter two estimates have been made under a "business as usual" scenario that does not consider the looming threats posed by global climate change or that effective future management may conserve more coral reefs.
- Forty-six percent of the world's reefs are regarded as being relatively healthy and not under any immediate threats of destruction, except from the "currently unpredictable" global climate threat.

Among the NOAA report's recommendations for action to conserve coral reefs are the following:

- Urgently combat global climate change
- Maximize coral reef resilience
- Scale up management of protected areas
- Include more reefs in marine protected areas (MPAs)

- Protect remote reefs
- Improve enforcement of MPA regulations; and
- Help improve decision making with better ecological and socioeconomic monitoring

The challenges facing coral reef health and survival today fall into two categories. The first category involves challenges that are more geographically localized due to specific human activities or practices. These include most of the threats or impacts discussed above: harmful coastal development, water pollution, sedimentation, overfishing or harmful fishing practices, and the effects of tourists, boats, and divers. The second category of challenges involves those that are global or regional in nature, the major ones being climate change (long-term and El Niño events) and associated ocean warming and coral bleaching and ocean acidification.

The first set of local impacts will need to be approached or attacked at a local level by the governmental agencies or legislative groups that control coastal land use and development practices, water quality and runoff, and fishing and related activities. The long-term benefits of healthy reefs to overall local economies and food supply have to be recognized and given priority over short-term profits for a few. While it is unlikely that intense development in areas of formerly healthy reefs, such as Singapore, Hong Kong, or Honolulu, will ever be removed (except by future sea-level rise), what is left can be preserved and protected if there is the political will to do so.

There is also cautious hope for reef rehabilitation. A team of researchers from Florida and Hawai'i has recently developed a process called microfragmentation to help barren, dead, or bleached sections of coral recover. The technique involves transplanting different species of coral tissue onto coral in land-based nursery settings. Early indications are that the growth rates are several times faster than those of their wild relatives offshore and offer some hope for restoring damaged reefs.

The larger global issues of a warming and more acidic ocean are far more complex. These will require global solutions to address and arrest climate change by wholesale changes in our use of energy, moving far more rapidly away from our dependence on fossil fuels and toward renewable energy resources. We have no other choice, and the sooner we accelerate that transition, the more likely it is that we will be able to save some of the planet's remaining coral reefs for ourselves, for our children and grandchildren, and for future generations.

# CHAPTER 15

# Fishing, Overfishing, and Aquaculture

Over 1.5 billion people, mostly in Asia, depend on fish for about 20 percent of their animal protein intake. Coastal waters are the most biologically productive on the planet, and about 99 percent of the annual worldwide commercial wild fish harvest comes from coastal waters within 200 nautical miles (220 statute miles) of a shoreline. Somewhere around 1,500 different types of fish are commercially fished in the world's oceans today, with the different stocks or species being fished or exploited to varying degrees. Overall, the abundance of perhaps one-third of these species is relatively well known because these are the fish that have been commercially harvested for decades. For the other two-thirds, and in many different geographic regions, the size of the remaining stock of fish is poorly known, and in some areas there are virtually no data on catch or populations.

A 2016 comprehensive global study on selected fish catch indicated that an estimated 30 percent of the total catch goes unreported. While the listed catch for 2010 was 85 million tons, careful study of individual fisheries revealed that about 35 million tons were unreported, making the total actually closer to 120 million tons for that year. It may be surprising that only about two-thirds of the fish caught today are actually consumed by humans; the other one-third is used for food for other animals, such as pigs, chickens, and farmed tuna and salmon, as well as for some industrial products.

While it would be extremely useful to know the standing stock of the world's different commercial fish species in order to begin to manage or plan for some level of sustainability, this is not likely to happen anytime soon. In large part this is due to the immense size of the world oceans, the fact that the fish are nearly invisible and always swimming, and the fact that so many individual nations are harvesting so many different kinds of fish in so many geographic areas.

## STATUS OF COMMERCIAL FISH STOCKS

Commercial fishing is a big business and also an important one, employing about 200 million people around the world. Although comprehensive data are lacking for all fish, we do have reasonably good numbers on the catch of the world's largest fisheries. In order by total commercial tonnage (2010), these include Peruvian anchovy, followed by Alaska pollock, skipjack tuna, Atlantic herring, Pacific mackerel, cutlassfish (a group of about 40 species of long, slender eel-like fish), European pilchard, Japanese anchovy, yellowtail tuna, and Atlantic cod. Perhaps not surprisingly, with the number of people on the planet growing at about 75 million every year (or 200,000 per day), the pressure on fish populations has continued to increase as well. The trend is clear, and the news is not good. Of the major oceanic fish stocks, 6 percent are classed as depleted, 3 percent are recovering, 16 percent are overexploited, 44 percent are considered fully or heavily exploited, 23 percent are moderately exploited, and just 9 percent are underexploited (figure 15.1). So over two-thirds are either depleted, overexploited, or being fished at their limit.

Overfishing and the intensive exploitation of the oceans have been taking place for years. The situation is steadily deteriorating, or rather it is slowly but steadily deteriorating. The total commercial catch peaked in about 1990 and has fluctuated between approximately 88 million and 95 million tons (as officially reported, but as mentioned above, these numbers are likely 30 percent too low) ever since (figure 15.2). The global commercial catch is about four times larger today than it was in 1950. Fish harvest numbers can be somewhat confusing, however, as the values reported by the U.N. Food and Agriculture Organization (FAO), which are used here, include the catches of not only fish but also other marine species such as prawns, mussels, and squid.

Despite declining fish stocks, fish catches have remained stable over the past twenty-five years or so because of a more intensified effort by the fishing fleets of the world. Boats have expanded into more distant areas

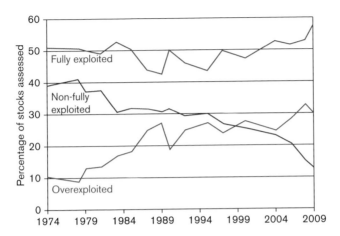

FIGURE 15.1. Status of global fisheries, 1974–2009.

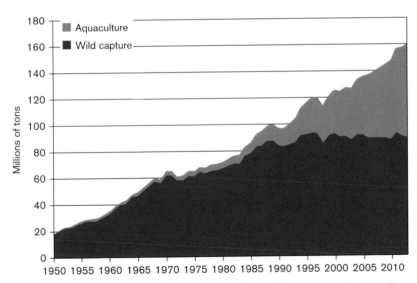

FIGURE 15.2. Global fish production, 1950–2013. The wild fish catch has leveled off while aquaculture production continues to increase. (Source: Food and Agriculture Organization of the United Nations)

far removed from the historical fishing grounds of the North Pacific and North Atlantic, with the southern oceans now being exploited. Fishing has also, like oil drilling, moved into deeper water. Whereas several decades ago fishing rarely occurred in water deeper than 1,500 feet, deep-sea trawls are now being towed at depths of up to 6,000 feet. And as soon as one species is exhausted, the industry typically turns to another previously unexploited species. Some of these newer targets have even been given new names in an effort to promote sales and make them more attractive to consumers.

A good example of this sequence of events is provided by the "slimehead," which once a fishery was initiated went on sale as "orange roughy." This was a fish that wasn't even on the radar screen until 1978, in large part because it inhabits cold and very deep water, 600 to 6,000 feet down. It is also a very slow growing and late to mature fish, which makes it very susceptible to overexploitation. The oldest orange roughies have been age dated through their otoliths, or ear bones, at about 150 years. From the earliest commercial fishing of roughies in the late 1970s, the fishery peaked in 1990 with a total global catch of 90,000 tons (figure 15.3). The harvest crashed quickly, and by 2009, just thirty years later, the total catch had plummeted to about 13,000 tons. Somewhat oddly, while this was the first commercial fish to appear on Australia's endangered species list due to overfishing, it has been neighboring New Zealand's highest-value fishery. Although orange roughy is still imported to the United States, it now appears on sustainable seafood guides, such as the Seafood Watch cards, with a red light warning not to consume.

In striking contrast to the very short, thirty-year slimehead or orange roughy fishery, the Atlantic cod is at the other end of the exploitation time line. The North Atlantic was fishing paradise for centuries, as cod became a highly desired and very abundant fish. Native Americans fished cod from the shore using simple hooks and nets. The first Europeans to harvest cod were the Basques and Vikings as early as 800 C.E. who found that drying or salting the fish could preserve them for later consumption. Shortly after the Pilgrims arrived in America in the early 1600s, they began to fish for cod, which started a new fishery that was to last for nearly four hundred years. There were so many codfish in the North Atlantic that they were thought to be inexhaustible.

By the early decades of the twentieth century, however, more effective fishing techniques were introduced, things like *gillnets* (vertical nets usually stretched out in a straight line for a considerable distance that catch fish by their gills, fins, or some other body part as they try to pass

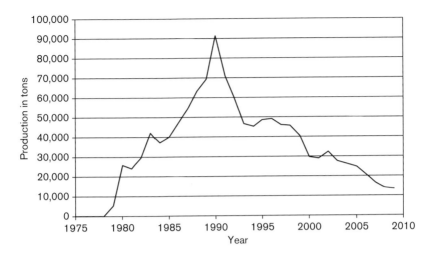

FIGURE 15.3. The rapid increase and then decline in the global production of orange roughy. (Source: Food and Agricultural Organization of the United Nations; image of orange roughy, Robbie Cada via Wikimedia)

through the nets) and *draggers* (another name for trawlers, which can pull a large trawl or net either along the bottom or at some specific depth), both of which led to large catch increases. Factory ships were also introduced that allowed them to freeze the cod at sea so the fishing boats could stay out longer.

The cod catch continued to increase throughout the twentieth century but peaked in the late 1960s, for both the Northwest and Northeast Atlantic stocks, with a combined catch of about 4.4 million tons (figure 15.4). With the passage of the Magnuson Fishery Act of 1976, foreign fishing boats could no longer enter the U.S. exclusive economic zone (EEZ), which extends 200 nautical miles from the shoreline. With this protection, the American cod fishery expanded further, but the cod were already in decline. Newfoundland had been codfish central for several centuries, and its harvest peaked with the rest of the North Atlantic in 1968, with a catch of 880,000 tons. Too many fishing boats,

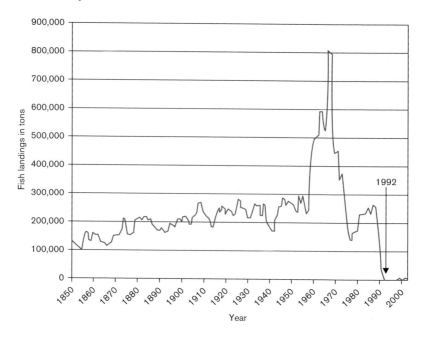

FIGURE 15.4. Historical trends in the catch of the Northwest stocks of Atlantic cod. (Image courtesy of Millennium Ecosystem Assessment; image of Atlantic cod, State of Massachusetts)

many of them foreign factory trawlers, all using modern technologies, chasing too few fish led to a huge crash in 1992. Catches dropped to 10 percent of their peak a little over twenty years earlier. In response, in 1992, after nearly five hundred years of fishing, the Canadian government declared a moratorium on the northern cod, the fish that had been at the center of eastern Canada's communities. Over 22,000 jobs were lost in over 400 coastal towns, but the total amount of cod was at 1 percent of its historic levels. There really was no other option.

The Canadian government responded thoughtfully, with income assistance and retraining programs for those who had lost their jobs. In the post-crash years, the economy and society of Newfoundland went through some major changes. The economy diversified, and with the decline in the predatory cod, snow crabs and northern shrimp populations expanded, so today there is a thriving new fishing industry that is approximately equal in value to the former cod fishery. An added benefit is that with the moratorium on cod fishing, cod populations are beginning to recover. The history and the governmental and civilian response provide important lessons to fishing communities around the planet. We have the power to maintain ecosystems and their living marine resources in order to protect the health and long-term sustainability of these large natural systems. It often just takes the political will.

The majestic and very large bluefin tuna is another sad example of an overexploited fish (figure 15.5). These fish are truly the raptors of the sea. They are hydrodynamically evolved to speed through the water at up to 25 miles per hour and dive to depths of half a mile. They are impressive fish that can grow up to 12 feet long and weigh up to three quarters of a ton. Their flesh is highly desired by the sushi eaters of the world as the finest available. As a result these fish have been pursued aggressively by large ships using spotter planes, caught in huge nets, and then often put in large ocean pens to fatten up prior to being butchered for steak and sushi markets in Japan, Europe, and the United States. Other bluefin tuna are caught by longlines, hook-and-line gear, and even harpoons.

In January 2014 a sushi restaurateur paid $70,000 for a 507-pound bluefin tuna. There are clear signs that these huge fish are in serious population decline, however, and major concerns that those in charge of managing fisheries are not doing enough to protect this species. Japan consumes about 80 percent of all the bluefin tuna caught worldwide as demand for their delicate and tender flesh has grown.

The populations of all three bluefin species, the Atlantic, Southern, and Pacific, have plummeted over the past fifteen years due to continued overfishing (see figure 15.5). And at $70,000 per fish, it's easy to see why bringing these fisheries under control is challenging. The stocks of the fish caught in the Atlantic and Mediterranean dropped by 60 percent in the decade between 1997 and 2007, and although there has been some improvement in recent years, the future prognosis is still highly questionable. Estimates are that over 90 percent of these fish are caught before they reach reproductive age, making sustainability very unlikely. An organization known as ICCAT (International Commission for the Conservation of Atlantic Tunas) is charged with regulating the bluefin tuna. According to some observers, ICCAT has now finally followed the recommendations of its own researchers.

FIGURE 15.5. A 1,000-pound bluefin tuna caught off Nova Scotia, Canada. (Photo: courtesy of Marc Towers)

The Pacific bluefin population is down to just 4 percent of its historic levels, spurring a recent symposium of two hundred policy, scientific, conservation, and industry leaders from twenty nations to discuss emerging science, management challenges, and the future of these fish. While there is a clear need to reduce the number of bluefin tuna being caught, fishery

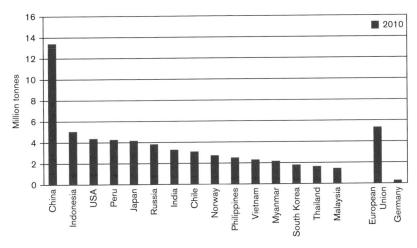

FIGURE 15.6. Global fish catch by nation, 2010. (Courtesy World Ocean Review)

managers haven't yet shown the political will to make the hard decisions necessary to protect this species. Two concepts that did rise to the top of approaches raised at the meeting were developing harvest strategy processes in the Atlantic and the Pacific (in essence, agreeing in advance on frameworks for management and management objectives) and using an electronic catch documentation system to track catch, trades, and illegal fishing. Getting the key players in the same room to talk about the problem at least is a positive step forward, but the difficult decisions and agreement on details still lie ahead. Talk is easy, and as has often been stated, after all was said and done a lot more was said than done.

## WHO IS CATCHING ALL THE FISH?

Using total tonnage landed by nation, at least as of 2010 when the last comprehensive FAO data are available, China was the most important fishing nation globally, and has been for years (figure 15.6). It reported a combined catch of 14.6 million tons in 2005, about 15.3 percent of the total catch. However, it is believed by many that the data are unreliable and that the tonnage had been adjusted upward to meet the central government's targets, although this may be changing. Interestingly, and adding to the challenge of documenting global fish catch, the recent study mentioned earlier indicates that global fish catch is underestimated by 30 percent, whereas China may be overestimating catch to meet national quotas.

Peru has been the second most important fishing nation historically but dropped to number four in recent years at 4.5 million tons landed annually. Its catch, which was dominated by Peruvian anchovies, declined 62 percent between 2006 and 2010 due at least in part to changing ocean conditions off the west coast of South America and as a result of a complete closure of the anchovy fishery to protect future stocks and let the fishery rebuild.

Indonesia is currently the second and the United States the third most important fishing nations with catches of 5.5 million tons and 4.6 million tons, respectively. Peru follows in fourth place, with Japan number 5, bringing in 4.4 million tons annually. Russia, India, Chile, Norway, and the Philippines follow in that order and round out the top ten fishing nations. The combined European Union catch, at 5.8 million tons, would rank number 2 in the world.

## FISHERY COLLAPSE AND RECOVERY

Fish populations have declined over the past fifty years, leaving many of the world's commercial fisheries in a challenging situation. The number of overexploited species has increased since the 1970s, and the total global catch peaked about 1996 and has been in a slow gradual decline ever since. One of the major driving forces during this period of heavy overfishing has been the Tragedy of the Commons in the sea, or the concept that natural resources like fish, without any clear ownership, will be harvested as long as possible until the resource has been completely exhausted.

As individual fisheries collapsed in the 1970s, 1980s, and 1990s, with the loss of thousands of jobs, it gradually became evident to both the fishing industry and policy makers around the world that overfishing was both an environmental and an economic problem. Some countries realized what had happened sooner than others and had the political will and power to make the changes necessary to stop the downward spiral. The United States, Canada, Australia, and New Zealand began to develop fisheries management plans that included catch limits for individual fisheries as one approach to allow stocks to recover to a sustainable level.

Europe has also gradually learned from some of its mistakes, although so many countries adjacent to the offshore waters of the North Sea and the Mediterranean made this more challenging. In the 1970s the overexploited herring fishery in the North Sea was completely closed for several years, which allowed the stock to recover. Accompanying the closure was a fisheries management program put in place to help prevent a future collapse. That is one positive example of sound fisheries management, but there are still many areas and fish stocks that are not being harvested with any measure of sustainability. The Mediterranean Sea is another area where there are a number of fish species that are being overexploited, in large part because it is surrounded by so many nations with fishing industries.

The global fisheries situation shows us examples of both successes and failures. Fish populations in areas where creative management has been used to develop sustainable fishing practices are recovering or have recovered so that fishing has again become part of the regional economy and harvests are increasing.

Working in favor of fish recovery is the ability of the females of many fish species to release thousands to millions of eggs each year. While most of these become food for other fish, at least they have evolved with the ability to recover under the right conditions for egg survival.

There are many other coastal areas, however, where competition for the remaining fish is very high and the Tragedy of the Commons principle is on full display. Short-term profits dominate over any long-term concerns, and stocks will in all likelihood continue to collapse until there is some intervention or the small catch returned is no longer justified by the costs. Some of the greatest challenges are (1) the difficulty of obtaining accurate or reliable stock assessments; (2) the lack of agreement between the fisheries scientists and the fishing industry on what is a reasonable or sustainable annual catch; and (3) natural population fluctuations because of changing oceanic conditions. The latter is common in the eastern Pacific where Pacific Decadal Oscillations produce major population shifts in anchovies, sardines, and squid. The ocean is a complex environment, and it is very difficult with natural year-to-year fluctuations in oceanic conditions (e.g., wind directions and strength, sea surface temperature, upwelling and nutrient availability), combined with fishery impacts from prior years, reproductive success, and species migrations with changing ocean conditions, to know just what to expect from one year to the next. The salmon and squid along California's coast can experience dramatic year-to-year population shifts related to both natural conditions and anthropogenic impacts.

MARINE PROTECTED AREAS

With the major decline or collapse of certain fish populations in recent years, proposals have been put forward to set aside fish sanctuaries or protected areas where fishing is not allowed or is tightly controlled in order to provide safe nursery grounds for young fish. In the United States over 10 percent of the total terrestrial land area has been set aside or preserved as parks, wildlife refuges, monuments, or wilderness areas, which have many benefits. In contrast, however, globally less than 2 percent of the ocean has been protected. Today the United States has established nearly 1,800 marine protected areas, which were chosen to protect spectacular reefs, tourist diving areas, underwater archaeological areas, and valuable fish habitats. These protected areas range from the Arctic to the tropics, and in size from one square mile to nearly 140,000 square miles.

In 1999 the State of California passed the Marine Life Protection Act, which was the first legislation in the nation that required the establishment of a science-based statewide system of offshore marine protected areas. The objective was simply to set aside and protect key habitat areas off the California coast with the objective of protecting and increasing fish stocks. The process of deciding which areas to protect and what type of protection to provide was a long and difficult but inclusive process that brought together a wide variety of stakeholders (including fishermen, conservationists, divers, native tribes, and local businesses), as well as scientists. The system of protected areas was completed in 2012 and the follow-on step was to set up a monitoring network to document how effective or successful these marine protected areas are.

The MPA design process started with some scientific guidelines, which were put forward as important criteria for successful sanctuaries and which included the following:

- *Size:* If the sanctuaries were too small, fish would simply leave or swim away, and if they were too big, fishermen argued, there wouldn't be enough areas left for fishing.
- *Spacing:* If the sanctuaries were too far apart, fish eggs or larvae would be carried by currents away from one protected area but would not reach the next.
- *Diversity of habitats:* It was deemed important to include a variety of habits, sandy bottom, rock outcrops, and kelp forests, for example.

The next step in the MPA process, which is still under way in California's coastal waters today, is to monitor these areas to see how effective

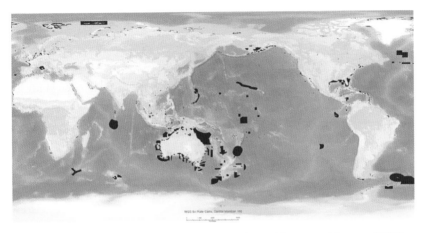

FIGURE 15.7. Marine Protected Areas throughout the world's oceans. (Courtesy IUCN and UNEP-WCMC, October 2013)

they are in allowing fish populations to expand and grow to maturity before being caught. Some fish grow slowly and only reproduce after fifteen or twenty years, so it will take some time before the benefits of these reserves are known. What has been documented to date is encouraging, however. The Marine Protected Area concept goes well beyond the United States, and there are now many areas around the world where critical fish habitat areas have been protected (figure 15.7).

## AQUACULTURE: FISH AND SHELLFISH FARMING

Simple fish farming has been practiced for hundreds of years, from carp ponds in early China to pre-Columbian fish traps in the Amazon Basin and coastal fishponds in the Hawaiian Islands. As wild fish stocks were depleted and fishing fleets had to go farther offshore to return a full catch, and in part to reduce the impact on the wild stocks and allow them to recover, fish farming has grown significantly as an industry in recent decades. The fish farming, or aquaculture, industry was relatively small, several million tons per year, until the mid-1980s, when it began to expand as the global wild caught harvest began to level off (see figure 15.2).

Between 2000 and 2010 the annual global aquaculture harvest doubled. By 2010 the production (including fish, shellfish, and crustaceans, raised in both salt water and freshwater) had expanded to about 75 million tons per year, with almost all of this in Asia (figure 15.8). Fish made

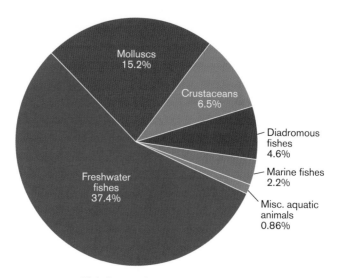

FIGURE 15.8. Global aquaculture production by type of catch, 2012. (Courtesy Food and Agriculture Organization of the United Nations)

up 50 percent of the total aquaculture harvest (2010), followed by aquatic plants with 23.75 percent, shellfish with 17.5 percent, and crustaceans with 8.75 percent of the total. While there are now over 200 species of marine animals and plants that are farmed, the major fish species are salmon, catfish, carp, Arctic char, trout, tilapia, and tuna. The shellfish are primarily mussels and oysters, and the crustaceans cultured or farmed include crabs, shrimp, and crayfish. In total abundance and market value, carp is number one, although there are a number of different carp species that are raised in many different countries. In 2013, 24.4 million tons of carp were harvested, all in freshwater, which produced a return of $31 billion. Atlantic salmon, which are raised in ocean pens, totaled 2.9 million tons and generated $10.1 billion. Tilapia, also raised in freshwater, was number 2 in tonnage at 3.6 million tons but was number 3 in income at $5.4 billion, still a large fishery.

China has become the world aquaculture leader, producing 61 percent of the total harvest, followed by Indonesia (8 percent), India (6 percent), Vietnam and the Philippines (each with 3 percent), and the Republic of Korea, Bangladesh, and Thailand (with 2 percent each) and Japan at 1 percent (figure 15.9). A number of other countries (Norway, Scotland, Spain, and the United States, for example) produce the remaining 12 percent.

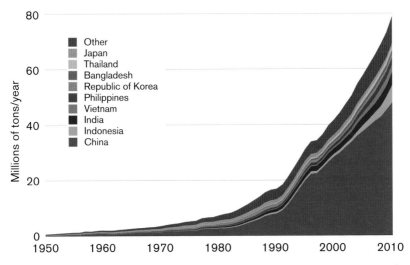

FIGURE 15.9. Global aquaculture production by country, 1950–2010. (Courtesy Food and Agriculture Organization of the United Nations)

While farmed fish and shellfish are providing an important supplement to our wild seafood supply, these farms cannot completely replace the diversity of wild fish and shellfish. Some aquaculture operations depend on wild populations to provide the eggs or the young that will be grown to maturity in the ponds. Many fish farming operations also still depend on certain species caught in the wild, such as anchovies and sardines, to provide food for the farmed species. Millions of tons of these smaller wild fish are caught in nets each year and then processed into fish oil and fishmeal for the farmed fish. Salmon, for example, which is one of the most important farmed species, is a carnivore and consumes over 3 pounds of fishmeal or wild fish to gain a pound in weight. Tuna, also carnivorous, must eat even more, about 15 pounds of feed for every pound it gains. So as presently practiced, fish farming has a large impact on wild fish populations.

Tilapia and catfish are two examples of farmed fish that are not carnivores so don't need to be fed fishmeal (figure 15.10). Raising these two species reduces the demand on the wild fish populations and, therefore, allows them to be grown at less cost and with less overall environmental impact. Growing mussels and oysters are two other good examples of sustainable aquaculture, where the shellfish feed by filtering food particles out of the water column. They can be grown from rafts, where

FIGURE 15.10. Tilapia farming is a major industry in Hong Kong, and because these fishes are herbivores, the environmental impacts are far less than for many other species. (Photo: Gary Griggs © 2014)

they attach to vertical arrays of wires or rope as they transition from their larval to shelled phase (figure 15.11).

BENEFITS OF AQUACULTURE

Without question, one of the major benefits of aquaculture is the ability to efficiently grow large quantities of fish, shellfish, or edible plants without having to expend the time and fuel hunting for fish in the ocean, sometimes hundreds of miles from your homeport. Aquaculture creates local jobs and can increase incomes at the community, regional, or national level. Fish and shellfish are an excellent source of protein, as well as essential amino acids, so can contribute to providing healthy food for increasing populations, particularly in Asia. This encourages local investment and can reduce the need to import seafood. Overall, the development of aquaculture can reduce fishing pressure on wild stocks or populations, ideally allowing those that may be overfished the opportunity to recover. It can also allow an alternative to harvesting fish or shellfish of questionable safety from coastal waters that may be

FIGURE 15.11. Mussels are cultivated in three dimensions in Spain, where they are grown on ropes in the water column suspended from surface floats. (Photo: D. Shrestha Ross © 2015)

polluted. These are all important benefits and have been among the reasons the aquaculture industry has grown rapidly in recent years and more opportunity exists in this field. Based on growth trends in recent years, it is likely that farmed fish and shellfish will soon surpass wild caught seafood.

## NEGATIVE IMPACTS OF AQUACULTURE

As aquaculture has expanded on a global scale, there have been some significant environmental impacts, and these issues need careful consideration before initiating any new aquaculture or fish farming enterprises anywhere. Efforts are under way to reduce or eliminate these impacts, but we still have a long way to go before these operations become sustainable and environmentally benign.

Salmon is one of the most valuable of all farmed fish; it is a popular seafood in many different countries, and as such it provides a good example of some of the negative environmental effects of aquaculture. Salmon consumption in the United States alone increased from 143,000 tons in 1989 to over 330,000 tons in 2004, more than doubling in fifteen years. Most of this increased consumption was due to the purchase of

FIGURE 15.12. Salmon farming in the lochs of Scotland is a huge industry, with most of the operations owned by Norwegian companies. (Photo: Gary Griggs © 2015)

farmed salmon, which is usually available at significantly lower prices than wild caught salmon. The wild salmon catch steadily declined throughout the twentieth century due to a combination of dams on major salmon streams and loss of spawning grounds, as well as changing ocean conditions. Today the great majority of salmon eaten by Americans are farm raised. Just over 2 million tons of Atlantic salmon, which is the most commonly farmed salmon, were harvested globally in 2013, with about one-third of that raised in the coastal waters of Norway and just under a third in Chile, Canada, Scotland, and the United States make up much of the remainder. Much of the salmon is raised in sea cages or sea pens, which are large enclosures, typically circular or square, placed just offshore and arranged in groups as a sea farm, connected by walkways (figure 15.12). These pens may be 100 feet in diameter and extend 30 feet below the surface, and each can house up to 90,000 fish.

Farmed salmon has been listed on the earliest Monterey Bay Aquarium Seafood Watch card as a fish to avoid due to an entire suite of environmental and health concerns. These include (1) parasites (e.g., sea lice), which are common when the fish are crowded into an enclosure; (2) diseases, which can spread rapidly in open net pens and can also be transferred to wild fish in the surrounding waters; (3) excrement, excess nutrients, and chemicals, which affect the surrounding waters and the seafloor below the pens; (4) use of antibiotics, anti-foulants, and pesticides and their impacts on other marine organisms; (5) concerns that farmed salmon, which are usually introduced from hatcheries, will escape from the pens and breed with wild salmon stocks; (6) dependency to date on fishmeal and fish oil, both primary ingredients in salmon feed, which puts additional pressure on wild fish stocks; and (7) excess nutrients from fish excrement and excess food, which can lead to algal growth, which ultimately decays and lowers the oxygen content of the water.

Some countries have been working to eliminate these environmental impacts by using recirculating water systems in order to reduce the impact on surrounding waters. Based on the aquaculture methods used, the Seafood Watch Card now designates some farmed salmon as the best choice (green, as in a green traffic light), some as a good alternative (yellow), and some to avoid (red) due to farming methods used. This program's rating system has guided consumers' choices in beneficial ways.

There are now other fish, catfish and tilapia, for example, that are becoming increasingly popular and are grown in closed freshwater systems, where wastes are controlled and fish cannot escape. These fish are herbivores and are fed a vegetable-based diet consisting of food made primarily from corn and soybeans, so there are no wild caught fish utilized for food. There are still some additional concerns with large-scale aquaculture, including the replacement of indigenous local fisheries with large multinational corporations. Most of the salmon farms now occupying many of the lochs in Scotland, for example, are actually owned by Norwegian companies.

## LOOKING TO THE FUTURE: WHERE DO WE GO FROM HERE?

Both wild caught and farmed seafood provide a significant portion of the protein supply for over 1.5 billion people, most of these in Asia. New fishing technologies have allowed the fishing industry to more easily locate and catch fish and locate and exploit areas that weren't harvested in the past. At present over two-thirds of ocean fish stocks are depleted, overexploited, or being fished at their limit. Global fish catches have been declining for the past thirty years or so, and this should be a wake-up call to all fishing nations that we have been removing most fish stocks far more rapidly then they can reproduce and grow to maturity.

International agreements on fishing quotas are critical if global fisheries are again going to be sustainable. These are extremely difficult to achieve, however, due to economic pressures, the lack of ownership and control of individual fisheries or fish stocks, and the difficulties of high seas enforcement. Yet without such agreements, each of these fisheries, especially those of species that require years to reach maturity and reproduce, will disappear or no longer be viable. Because many prime spawning grounds and nurseries lie in coastal waters controlled by individual countries, there is hope for the future through conservation efforts by these nations and also through the establishment of more marine protected areas. Political will and enforcement are necessary in both cases.

The decline in ocean fish catch over the past several decades as a result of overfishing practices has been compensated for, however, by substantial increases in farmed fish and shellfish. The global aquaculture harvest is now approaching the annual wild catch, and it is highly likely that this trend will continue, with fish farming soon overtaking hunting for fish in offshore waters. While this expanding land-based industry is providing an important source of protein, employment, and income, particularly in Asia, where the great majority of the world's aquaculture is now concentrated, it is not without its problems. These include concentrated wastes and pollution from pen crowding, spread of endemic diseases, high concentrations of pathogens and parasites, and increased pressure on some wild stocks to provide food for farmed fish. Awareness and publicity of these impacts is leading to consumers making better choices and also providing incentives to improve aquaculture practices to reduce or eliminate the negative impacts.

# CHAPTER 16

# Aquatic Invasive Species

Aquatic invasive species (also called exotic or alien species) are animals or plants that evolved in one region or area and are either accidentally or intentionally introduced through various means into another area. Because these introduced species often lack natural population controls such as predators in their new home, they can cause major ecological disruption or instability by outcompeting and overwhelming the native populations. They may also introduce new diseases and/or parasites. There are a number of instances where these invaders have completely transformed natural ecosystems. Over seven thousand species are known to have been introduced into the United States alone, of which about 15 percent have caused ecological damage and economic losses to agriculture, fisheries, and forestry, as well as human health. These invasive species cause an estimated $137 billion in losses and damage in the United States each year.

To be fair, however, there are also large numbers of commercially valuable species, particularly agricultural plants, that have been transplanted from other parts of the world to the United States. In the case of California, they have surrounded us for well over a century and are now the backbones of California's agricultural economy. Many of these were intentionally brought from Central America, Asia, and Europe and include citrus, avocados, olives, figs, and artichokes, to name just a few. Honeybees are also alien, having been brought here from Asia, and the agriculture industry relies heavily on them. Then, on the flip side, there are a host of other species in California that are so invasive and ubiquitous that many residents are probably not aware that they were even introduced from elsewhere: Scotch and French broom, acacia, eucalyptus, pampas grass, ice plant, poison hemlock, and thistle, to name a few in coastal California.

## INVADERS FROM THE SEA

Animals have always used the oceans to move about the planet. By swimming or hitching a ride on a log, branch, or coconut, organisms have found new worlds in which to thrive. But until recently this process has proceeded at a geologic pace, often limited by winds, currents, and the large expanse of ocean between continents or islands. Once humans figured out how to build boats, and now massive ships, however, intrepid stowaways have had an ever-expanding group of vehicles for dispersing themselves both faster and farther. The result is an increasing number of ocean ecosystems, primarily near coastlines, whose inhabitants are being threatened, eaten, displaced, or outcompeted by non-native species.

Most invasive marine species stow away in ballast water. Many large ships have tanks in their hulls that are filled with seawater to provide stability. The water is pumped into ballast tanks when the cargo is being unloaded, and discharged when the cargo is being loaded (figure 16.1). In the case of oil tankers, the cargo provides the ballast, but after it is offloaded, seawater is typically taken on to provide stability as they make their return trip. Ships may take on 10 million to 20 million gallons of ocean water. While at sea, ships may also adjust the ballast water load in order to compensate for fuel consumption, waves, and weather. The ballast water taken on in port will include everything living in the ocean at that particular location that passes through the intake pipes. When the ship arrives at its destination, it pumps out the ballast water—along with whatever species happen to have been along for the voyage, from microscopic organisms, eggs, and larvae of a wide variety of marine life to entire schools of small fish. Today, with approximately 95 percent of our international exports and imports shipped by sea, there is ample opportunity for hitchhiking.

Some 45,000 cargo ships move an estimated 10 billion tons of ballast water around the world every year. In addition to ballast water, many organisms attach themselves to the hulls of ships (known as marine fouling) and on the millions of tons of plastics and other trash that float around the globe in ocean currents. Pets acquired through the aquarium

FIGURE 16.1. Large containership pumping out ballast water. (Photo: Greg Goebel licensed under CC BY 2.0 via Flickr)

and exotic pet trade—and then released—can become invasive species, as can escapees from aquaculture farms. And the ongoing ocean temperature rise caused by global climate change is allowing non-native species to populate ocean habitats that were once too cold to be hospitable.

In early June 2012, a 66-foot-long concrete and steel floating dock washed onto the central Oregon coast near Agate Beach (figure 16.2). The Japanese consulate in Portland confirmed that the dock was one of four used by commercial fishermen for unloading squid and other seafood catch at the port of Misawa that had been ripped away from the coast during the March 2011 tsunami. It took about fifteen months for the floating structure to make the roughly 5,000-mile trip across the North Pacific, traveling about 10 miles a day. Scientists from Oregon State University's Hatfield Marine Science Center discovered that the dock contained an estimated 100 tons of encrusting organisms, or about 13 pounds per square foot. These included several species of barnacles, as well as mussels, sea stars, urchins, anemones, worms, limpets, snails, and algae—in short, dozens of species.

Although most of the individual species were unique to Asia, this smorgasbord of marine organisms was similar to what you might find on a wharf or piling along the coast of California, Oregon, or Washington. The State Department of Fish and Wildlife set to work scraping, bagging, and then burying all of the organisms in order to minimize the potential spread of non-native species. But they also were clear in pointing out that there is no way of knowing if any of the hitchhiking

FIGURE 16.2. Following the 2011 tsunami, a dock from Japan arrived on the Oregon coast. The state Department of Fish and Wildlife scraped all attached marine life into buckets and buried it. (Chris Eardley, NOAA/NGDC /Oregon Dept. of Fish & Wildlife)

organisms or their eggs or larvae had already jumped ship and headed for new homes along the West Coast. So while the Oregon Department of Fish and Wildlife was attempting to be very thorough in its systematic annihilation of every living organism on that floating dock, the cat may already have been out of the bag for some time.

While the initial arrival of Japanese tsunami debris briefly brought the issue of invasive or introduced marine species into a more public light, introduced terrestrial plants and animals have surrounded us for well over a century. With so much of our imports arriving by sea, whether the larvae or eggs are discharged in ballast water or came from a bit of tsunami debris, they are all invasive species, some of which present serious problems and all of which can be very difficult to control.

## THE INVASION OF SAN FRANCISCO BAY

Invasive species is not a new issue. In the San Francisco Bay delta, the problem dates to at least the California gold rush, when barrels of eastern oysters were shipped west to San Francisco aboard transcontinental

FIGURE 16.3. Containership from Asia entering San Francisco Bay. (Photo: Willem van Valkenburg licensed under CC BY 2.0 via Flickr)

trains. Some of them, along with the eastern seaweeds they were packed in, found their way into bay waters and proliferated. The problem got progressively worse over the years, as San Francisco Bay and its wharves, docks, and navigation channels became a center for global shipping for the western United States. Today an estimated 7,000 containerships and about 10,000 oil tankers call at ports in the bay and delta every year (figure 16.3). Each ship may contain 10 million to 20 million gallons of ballast water, pumped on board from some foreign port. While the ballast water keeps a ship stable when at sea, the water discharged on reaching port contains organisms and eggs from those distant waters, which may become a huge ecological destabilizer.

As a result, San Francisco Bay is now often recognized as the most invaded waterway in the world. More than two hundred animal and plant species are reported to have taken up new homes and are thriving in the waters from the offshore Farallon Islands to the inland port of Sacramento. Fifty-six of these are considered to be harmful. Asian clams, Chinese mitten crabs, African clawed frogs, Amur River clams, New Zealand carnivorous sea slugs, Black Sea jellyfish, and Japanese gobies are just a few of the exotics that now inhabit the bay and its adjacent waters (figure 16.4). In individual parts of the bay-delta complex, invasive species may make up 40 to 100 percent of the common species and up to a surprising 97 percent of the total number of organisms.

The problem in San Francisco Bay isn't diminishing. Between 1851 and 1960, there was only about one new introduced species recognized

FIGURE 16.4. Some of most problematic invasive species in San Francisco Bay, California: (a) Chinese mitten crab (Photo: AFPMB licensed under CC BY-NC-ND 2.0 via Flickr); (b) New Zealand mud snail (Photo: courtesy of Penn State University); (c) African clawed frog (Photo: Brian Gratwicke licensed under CC BY 2.0 via Flickr)

(C)

FIGURE 16.4. *(continued)*

each year. From 1961 to 1995, however, as ship traffic increased, invasive species arrived at the rate of about one every three and a half months.

In 1998 a ten-year-old girl discovered a small creature in a Palo Alto creek that flows into South San Francisco Bay that could have been out of a science fiction horror movie. It looked like a large tarantula but with hairy claws and long spiny legs protruding from a dark shell. About the same time, a thirteen-year-old boy found similar creatures in a drainage canal over 100 miles up the Sacramento River from the Golden Gate, near Rio Linda, and began catching them in buckets to show his friends. Two years earlier, in 1996, biologists found 45 of these exotic creatures trapped on fish screens at water pumps in the Sacramento delta. In 1998 they returned and found 25,000 caught on the intake screens in a single day.

This bizarre creature, the Chinese mitten crab, has turned thousands of miles of California waterways and canals into a bad movie (figure 16.4). They have shown up nearly everywhere in Central California, from San Francisco Bay to streams in the Sierra foothills. The mitten crab, named after the dense patches of hair on its claws that resemble mittens, is native to the coastal rivers and estuaries of China. It spread through Europe in the early part of the past century and apparently was first discovered in California in 1993 in San Francisco Bay. It isn't clear whether these crabs came in ballast waters or were illegally introduced. The latter is considered highly likely, as there is a lucrative market for them in China, where they sell for $10 to $20 apiece, prized because the crab ovaries are believed to provide magical powers after being consumed.

Imports and sales were banned in California by the Department of Fish and Wildlife after it was learned that live crabs were being sold in both San Francisco and Los Angeles. The species can carry a lung fluke, and instead of bestowing magical powers, they can cause symptoms very much like tuberculosis that can infect anyone eating raw or incompletely cooked crabs. The crab population has exploded, however, and has presented California with very some serious problems. These hairy crabs can devour a variety of bottom-dwelling animals, from shrimp to young shad and potentially eggs and juveniles of salmon and sturgeon. The crabs burrow into stream banks and levees, which can accelerate erosion and reduce levee stability and safety. They have repeatedly clogged water intake structures in large numbers. California could well have done without the Chinese mitten crab, but it's now going to be very difficult to get rid of it.

## CONCERNS WITH INVASIVE SPECIES

In contrast to the negative impacts of mitten crabs in Central California, there are actually a few foreign marine species that are being considered for introduction because of their perceived benefits. Asian oysters, for example, are better at filtering out water pollutants than native oysters. They also are more resistant to disease and grow faster than the natives. Biologists are currently considering introducing the oysters into Chesapeake Bay to help restore stocks and remove pollutants. Any effort of this sort, however, needs to be studied very carefully in a controlled situation so that all aspects of the species' life cycle are understood before introduction to waters in the United States.

More commonly, invasive or exotic species become problematic because they often grow fast, reproduce rapidly, have the ability to disperse widely, may have a tolerance for a wide range of environmental conditions, can live off of a wide range of food types, and can displace or outcompete the native populations, whether flora or fauna. They may also prey or become parasitic on native species and transmit diseases, or they may even have an impact on human health. Some examples of the direct impacts of these invaders are clogging of navigable shipping channels and canals, clogging water intake screens or inlets, and reducing commercial or sport fishing populations of fish and

FIGURE 16.5. *Caulerpa taxifolia* is a highly invasive algae. (Photo: Kyle Demes/ Smithsonian Tropical Research Institute)

shellfish. Damage can be extensive, with estimated losses and control costs for all invasive species in California, terrestrial and marine, being about $3 billion annually.

In June 2000 several scientific divers went for what they thought was a routine research swim through an eelgrass bed in Agua Hedionda Lagoon in northern San Diego County, California. Working on a restoration project, they were swimming transects, measuring the extent of the eelgrass bed and noting new shoots. Then one of the divers came face-to-face with a large patch of an unusually green, beautiful feathery seaweed. This strange plant would later be identified as the first confirmed North American sighting of *Caulerpa*, what has been called "the killer algae" (figure 16.5).

It is normally native to tropical waters in the Indian Ocean and is used as a decorative seaweed in saltwater aquaria because it is hardy, fast growing, and not eaten by fish. However, when introduced into a suitable new habitat (typically when people naively dump their aquarium into some local water body), it quickly becomes a dominant and persistent species that displaces native seaweeds and crowds out other marine life.

While not a combative "killer" in the true sense of the word, this renegade aquarium plant grows rapidly and can form a smothering blanket over mud, sand, or rock, severely reducing native populations of seaweed and seagrass. It can take over the natural eelgrass beds, which provide habitat for lobsters, flatfish, and bass, threatening their populations. It also can spread or grow up to ten times as fast as the native seagrass. It has been found in this one lagoon in San Diego County but also in Huntington Harbor in Orange County. Eradication efforts were quickly initiated through a process of covering the *Caulerpa* with tarps and chlorination of the affected areas, which was successful, at least until the next aquarium is innocently dumped out somewhere. This plant can reproduce from small fragments, so a lot of care and vigilance is required to eliminate all occurrences. It was also discovered several years ago near Sydney, Australia.

In 1984 a cold water–tolerant variety of *Caulerpa* developed for the aquarium industry was first introduced into the Mediterranean Sea, either accidentally or intentionally, where it overwhelmed aquatic ecosystems. In ten years it had covered a reported 7,400 acres, and it continued to spread, preventing native plants from growing. There has been considerable disagreement, however, on both the extent and the impact of this alien algae. It has been successful in reducing pollutants, principally nutrients from sewage disposal, and while there have been claims that over half of local fish species have disappeared, scientific studies have shown that diversity and amount of fish are equal or greater in *Caulerpa* areas than in seagrass beds. There have also been statements that the areas colonized or invaded by *Caulerpa* have been greatly exaggerated. These contrasting statements and assessments point to some of the challenges of finding unequivocal conclusions about many environmental impacts and issues of this sort.

Another invader on a larger geographic scale is the zebra mussel, accidentally introduced by a cargo ship into the North American Great Lakes from the Black Sea in 1988. The tiny mollusk multiplied uncontrollably, starving out many of the Great Lakes' native mussel populations and thereby negatively affecting commercial and recreational fisheries. The small mussel also interferes with the intake pipes for city water supplies as well as factories and also fouls ship rudders. This nasty little invader has now spread from Canada to Mexico and is considered a major nuisance species. Hundreds of millions of dollars are spent annually in attempts to control its numbers. Although zebra mussels are exceedingly hardy, scientists are working to identify predators, parasites, and infectious microbes that can kill them but will leave native shellfish unharmed. This, however, is always a tricky proposition, simply because the impacts of introducing yet another non-native specices are not always clear at the outset.

The Hawaiian Islands, originally an isolated and ecologically unique habitat, now receive ships from all over the world, and also host a major U.S. Navy base at Pearl Harbor. The marine traffic and its ballast water and hitchhikers have had major impacts on the islands' unique flora and fauna. The snowflake coral is believed to have been originally introduced to the islands as larvae in ballast water or on the bottom of a ship as hull fouling. It was first discovered in Hawai'i in 1972 at Pearl Harbor and has now spread to all the major Hawaiian Islands. It grows well on both natural and artificial surfaces such as metal, concrete, plastic, and even rope. It has become a highly successful invader and threatens Hawai'i's biodiversity by monopolizing food and space resources and by displacing native species. It has been observed growing on and smothering black coral colonies at an astonishing rate, and it appears to have no significant predators in Hawai'i to limit its expansion.

Islands overall represent the greatest concentration of both biodiversity and species diversity on the planet. Species living on islands are typically unique, but they are also highly vulnerable to disturbance and predation. Although less than 5 percent of the Earth's landmass consists of islands, 80 percent of known extinctions over the past five hundred years have taken place on islands, and 40 percent of the animals around the world currently at risk of extinction occur on islands. Globally islands have been heavily affected by the human introduction of non-native species; some of the most problematic, such as goats and pigs, were intentionally introduced or were brought by humans for a food supply. Some became feral (cats), or simply accompanied humans (rats).

A dedicated group of wildlife biologists first recognized this problem over twenty years ago, initially on the barren rocky islands in the Gulf of California. Island Conservation, an environmental nonprofit organization, grew out of the initial efforts and began to work with local communities, government management agencies, and conservation organizations to select islands that had the greatest opportunity for preventing species extinction and then developed plans to remove the invasive species. By removing the feral animals and incrementally fencing off areas, seabirds soon returned to nest and began to fertilize the ground, which was followed by the gradual recolonization by native plants. This has been a remarkably successful grassroots effort that has now protected 994 populations of 389 different species on 53 islands all over the world.

## THE GLOBAL THREAT AND COMBATING POTENTIAL INVASIVE SPECIES

An estimated 84 percent of the world's coasts are being affected by foreign aquatic species from somewhere, and there are very few pristine coastlines left on the planet today that are immune from invasives. Among the environments that have been most heavily affected are the major global ports. A recent study utilized a number of different factors to determine which global ports are at the highest risks of invasive species. The key variables appear to be environmental conditions such as water temperature, marine biogeography or the distribution of marine plants and animals, and ballast water release restrictions or protocols. There also appears to be a "sweet spot" that influences the probability of a foreign species surviving in a new location, which turns out to be about 5,000 to 6,000 miles from the original home or source. At shorter distances it is unlikely that the organisms being transported and introduced are truly foreign to that location; at longer distances the odds of surviving the voyage become significantly reduced.

The study concluded that the world port at highest risk of invasive species is Singapore, probably because of its location in the middle of major shipping routes and because it is one of the world's busiest ports and has reasonably warm water (figure 16.6). Hong Kong, another major port, followed Singapore. Most of the other ports or locations most likely to be attacked by invasive species are in Asia, the Middle East, and the United States. These include the Suez and Panama Canals, Kaohsiung (Taiwan), Port Said (Egypt), Busan (Korea), Jebel Ali (UAE), Kawasaki (Japan), Durban (South Africa), Yokohama (Japan), and then New York, New Jersey, and Long Angeles–Long Beach in the United States.

In an ideal world, the best and most cost-effective approach to the threat posed by invasion of potentially harmful alien species is to prevent them from setting up house in the first place rather than trying to eradicate them once they have been established. This approach requires an understanding of the routes or ways in which invasive species move from point A to point B. The consensus seems to be that ballast water is the main transport mechanism, so attacking this process is key to reducing further invasions.

FIGURE 16.6. Ships anchored offshore at Singapore, one of the world's busiest ports. (Photo: D. Shrestha Ross © 2014)

In order to combat invasive species, some governments are focusing on how they handle ballast water. New regulations in a handful of countries require ships to exchange their ballast water before reaching port. This releases the organisms, eggs, or larvae from their original port into the open ocean well before arriving at their destination port, where they are less likely to survive. The ship then takes on ocean water for ballast with its own collection of plankton that will be released in the destination port, but these are not believed likely to be harmful or survive long or establish themselves in port conditions. This is certainly a promising approach, although it may not always be feasible. During large storms with high winds and waves, ballast water exchange could lead to an unstable and unsafe ship. A complementary approach is the development of methods to treat the ballast water in order to kill the organisms that are contained as passengers or stowaways.

Hawai'i, because of its vulnerability to invasive species, has been a leader in developing approaches to reduce the potential for introduction of new marine species. In 2000 the state's Department of Land and Natural Resources was designated the lead agency for prevention and eradication of alien aquatic organisms that could be introduced through either ballast water or hull attachment. They have created the Alien Aquatic Organism Task Force, with membership that includes federal and state agencies, private shipping and boating industries, environmental organizations, and the scientific community. In 2007, after extensive consultation, discussions, and public input, the task force adopted a set of rules for managing ballast water discharge from ships in Hawaiian waters. The rules were developed to minimize the introduction and spread of nonindigenous aquatic organisms in waters surrounding the Hawaiian Islands. They require that each ship has a ballast water management plan, that it files a ballast water reporting form with the Department of Land and Natural Resources twenty-four hours prior to arrival in Hawai'i, and that it follows all of the rules contained in the adopted guide.

## WHERE DO WE GO FROM HERE?

We have a truly global economy with nearly every country on the planet exporting and importing goods, the great bulk of these moved by ships, which carry and discharge ballast water. In 2015 the total U.S. trade with foreign countries was $4.99 trillion, which includes $2.23 trillion in exports and $2.76 trillion in imports, making the United States the world's largest importer and the third largest exporter, after China and the European Union. The United States imports about 40 to 45 percent of our oil, some 9 million to 10 million barrels every day, much of which comes into a number of American ports by tanker.

Ships with marine organisms attached to their hulls and riding along with ballast water will continue to be transported around the world's oceans. Although research is under way to determine the effectiveness of treating ballast water to disinfect it before discharge, there is a long way to go before this is a required or common practice. Similarly, regulations or rules requiring exchange of ballast water at sea are still in their infancy and are not yet universally required.

Cargo ships have recently begun to carry automatic identification systems that track their movements, giving researchers new insight into the vessels' journeys. This may be helpful in determining which ships and routes we need to pay closer attention to.

Despite these efforts, it is likely that continued changes in the marine environment at both ends of a ship's voyage would alter both the organisms that might be transported and the degree of susceptibility to invasions. The potential for additional species to take up new homes will no doubt continue to be an issue for the immediate future. Regardless of what future steps we take to reduce the number of new alien species being introduced, we still have a large number of species that have already become established in their new homes. With this in mind, another approach is to evaluate invasive species that have already become established to see if they can be contained if not eradicated. International regulations requiring ballast water exchange prior to port arrival as well as treatment of ballast water to kill hitchhiking organisms appear to be our best approaches for controlling this global issue.

CHAPTER 17

# Sand, Dams, and Beaches

If you stop to think about it for just a minute, the fact that beaches exist as a long, nearly continuous ribbon around much of the world's 372,000 miles of coastlines is pretty amazing. The United States alone has nearly 11,000 miles of beaches. What would the coast of California or Hawai'i be like without them? Perhaps one long highway filled with cars, parking lots, and overlooks for a brief view of the coastline and ocean. These beaches provide us all with a multitude of opportunities, swimming, surfing, jogging, volleyball and soccer, walking and beachcombing, or just sitting and relaxing; the list is almost endless, which is why millions of people visit them each year. In fact, people all over the world spend thousands of dollars to fly or drive to warm sunny places with beaches for their vacations. Most northern Europeans flock to the Mediterranean beaches in late summer, filling up the highways, campgrounds, hotels, and beaches. Florida, Southern California, the Mexican Riviera, the Caribbean Islands, and Hawai'i all play a similar role in drawing people to the shoreline to rest, relax, and recuperate from the stresses of their normal lives (figure 17.1).

Sandy beaches only exist, however, because of a unique combination of mineral properties and wave energy. While there are beaches around the planet that consist of broken coral and shells and others that are made up of flat pebbles known as shingles, or even larger cobbles, by and large the great majority of the world's beaches are made up of sand. It's the Goldilocks material, just the right size. Finer-grained silt and

FIGURE 17.1. A crowded July 4th weekend at Main Beach in Santa Cruz on the Central California coast. (Photo: Gary Griggs © 2015)

clay do not remain on beaches because breaking waves and coastal currents carry them offshore. Pebbles and cobbles are usually too coarse and heavy for waves to carry them very far, so they tend to accumulate close to their source area, either at the base of a rocky seacliff or near river mouths where they were delivered to the shoreline. Cobble or pebble beaches are generally not the places where people spread their towels, play volleyball, or take long, quiet walks, however. It's just not that comfortable lying or walking on angular shells or baseball-size cobbles.

The ideal beach for most of us consists of fine- to medium-grained, clean, white sand. What is interesting and what makes most beaches around the world possible is that many rocks, granite, for example, consist of a group of minerals that tend to break down over time into sand-sized particles. Either through physical or chemical weathering in the watersheds or stream channels, or along the shoreline, these rocks decompose and wear down into small, individual mineral grains; quartz and feldspar are among the most abundant and also the hardest and therefore most durable. Rivers can easily transport sand downstream toward the shoreline, and waves can then take over and continue to sort

FIGURE 17.2. A steep gravel beach on the southwestern coast of England at the village of Beer. (Photo: Gary Griggs © 2015)

and round the grains until they are perfect to walk on. What if most rocks instead broke down into marble- or golf ball–sized fragments, and this was what we found on most beaches? We do find beaches like this, and they are interesting and different, but they are not usually covered with people (figure 17.2).

These narrow sandy strips provide some enormous benefits, yet they are under constant threat. Not only does sand face a lot of obstacles getting to the shoreline, but it must overcome challenges to remain there.

## RIVERS, SEDIMENTS, DAMS, AND BEACH SAND

Beach sand can come from several different sources depending on the adjacent terrestrial topography, geology, and climate. In California, which is typical of most of the U.S. West Coast, the great majority of sand is delivered by rivers and streams, with lesser amounts contributed by bluff or cliff erosion. California, Oregon, and Washington are fortunate in having many rivers that deliver large quantities of sand to the coastline to nourish their beaches.

Rivers transport the bulk of their sand during times of high flow. In fact, it is the occasional floods or very high discharges, perhaps during just a day or two each year, that carry by far the great majority of sediment. A few days of very high river flow can make up for ten or more years of low flow when it comes to sand delivery. Years with little rainfall and low river flows do not contribute much sand at all to the beaches, but a big flood can make up for many drier years. In 1969 over 100 million tons of sediment (about 7.5 million dump truck loads) was flushed out of the Santa Ynez Mountains in Southern California and onto the shoreline. That was more than the previous twenty-five years combined. Even more amazing, in a single day during a large flood on December 23, 1964, the Eel River in Northern California transported 57 million tons of sediment (about 4.2 million dump truck loads), which represented nearly 20 percent of the sediment load carried by the river over the previous ten years.

All is not well in the watersheds of the world, however, when it comes to beach sand supply. In California the steeper upstream reaches of many rivers have been dammed for some combination of water supply, flood control, recreation, and hydroelectric power generation. Their broad, low-relief, downstream alluvial channels are often mined for sand and gravel for the construction industry, and in urban areas they have often been converted to concrete-lined channels to control flow. Each of these reduces or eliminates the sand that was transported downstream under previous natural conditions. Over the past 125 years, California has built nearly five hundred dams on coastal rivers and streams that formerly flowed freely to the coastline (figure 17.3). In addition to their benefits, large dams have two major negative impacts on sand transport to the shoreline: (1) they reduce or control the large flood flows that transport the great majority of the sand, and (2) they trap nearly all of the sand being carried by the stream in their reservoirs and prevent it from reaching the shoreline and the beaches (figure 17.4).

Under natural conditions, California's coastal rivers, large and small, delivered on average about 13 million cubic yards of sand to the coastline each year. The construction of the hundreds of dams reduced this beach sand delivery by about 25 percent, however, and those dams have now trapped or impounded about 200 million cubic yards of sand. That's equivalent to about 20 million dump truck loads of sand, enough to build a beach 100 feet wide and 10 feet deep along nearly the entire 1,100-mile length of California's coastline. Although sand impoundment was not an intended consequence of dam construction, today we

Background map by Raven Maps and Images

FIGURE 17.3. California has built nearly 500 dams on the state's rivers and streams. (Courtesy: California Division of Safety of Dams and Raven Maps and Images)

understand that this is a predictable outcome of these river barriers. The benefits of flood control, increased water supply, recreation, and, in some cases, hydroelectric power generation (which is clean and renewable) have been countered to some degree by the gradual reduction of sand input to the shoreline from these dammed rivers, as well as the

FIGURE 17.4. This reservoir behind San Clemente Dam on the Carmel River, Central California, contained 80 percent sediment before being demolished (Photo: Gary Griggs © 2009).

disruption of the migration of important anadromous fish such as salmon and steelhead trout.

California is not alone in building dams. Across the United States there are now approximately 75,000 dams, impounding 600,000 miles of river, which is equivalent to almost one dam being built every day since the Declaration of Independence was signed in 1776. Around the planet there are an estimated 845,000 dams. China actually had over 22,000 dams (over 50 feet in height) as of 2000 and has built about 20 percent of the world's large dams, almost half of which are for irrigation. When their present plans for additional dams are completed, there will not be a single river in the country that has not been dammed.

The Three Gorges Dam on the Yangtze River created the world's largest reservoir and also has the world's largest power station, although its construction didn't come without considerable controversy and opposition (figure 17.5). While sand impoundment has been a big impact of dams in California, in China one of the larger initial arguments against the construction of the Three Gorges Dam was the inundation of 13 cities, 140 towns, and 1,350 villages that would accompany reservoir

FIGURE 17.5. The Three Gorges Dam on the Yangtze River, completed in 2006, is the second largest hydroelectric-generating dam in the world and impounds a reservoir that extends over 400 miles upstream. (Image: putneymark licensed under CC BY-SA 2.0 via Flickr)

filling and that required resettlement of at least 1.24 million people. Now that it has been completed, the massive dam has a number of acknowledged social, environmental, ecological, and geologic problems. You can't build a dam that large and impound a reservoir that long and not expect some significant environmental effects. The watershed is highly erodible, and about 40 millions tons of sediment are now being washed into and deposited in the reservoir each year. This sediment is no longer nourishing downstream floodplains, and it is not helping to build out a delta at the shoreline or nourish the shoreline with sand.

The Aswan Dam on the Upper Nile is the Three Gorges Dam of Egypt (figure 17.6). It is another example of the incompletely understood direct and indirect consequences of constructing a very large dam. The objectives were similar to all large dams: to produce electricity, control floods, and store water for irrigation that would allow for the expansion of agricultural production along the Nile River. But the balance of benefits and costs remains controversial years after its completion. From the upstream reservoir to the downstream Nile Delta and the eastern Mediterranean, the impacts were numerous and troubling and could have been anticipated. The trapping of the Nile's sediment load behind the dam was one of the major problems and has had a number of direct effects. Deprived of the Nile's rich sediment, much of the downstream farmland now needs fertilizer. Two-thirds of the annual

FIGURE 17.6. The Aswan Dam on the Upper Nile, Egypt, impounds Lake Nasser, which extends 420 miles upstream. (Imagery © 2016 DigitalGlobe, DigitalGlobe, CNES/Astrium, Map data © 2016)

fertilizer usage is needed to make up for the lost fertility and mineral content since the Nile silt stopped coming. Because of the decline in sediment discharge and therefore delta outbuilding, the Mediterranean Sea is eroding the delta. This area is home to some of Egypt's most productive farmland and is being lost at estimated rates of 400 to 500 feet per year. Nutrient delivery to the Mediterranean was also reduced, and is believed to have been a major factor in the decline in fish catch, although this is a complicated issue. The former sediment load of the river, about 134 million tons per year, is now deposited in the reservoir and is slowly reducing the water storage capacity of Lake Nasser. These are some of the major effects of the Aswan Dam trapping large volumes of sediment, which for centuries provided the lifeblood of Egypt's agricultural productivity.

While it is possible to build dams with outlet pipes or sluices that could be used to let sediment wash out from the bottom of a reservoir, this wasn't a concern or issue when any of the existing dams were built. Dam builders were focused on the benefits, and if loss of sediment was ever considered, it was downplayed. So we are stuck with hundreds of millions of tons of sediment sitting at the bottom of thousands of reservoirs around the world that could be adding fertile silt to floodplains and deltas and providing sand to nourish beaches. While the attraction and benefits of constructing additional large dams for both clean hydro-electric power and irrigation water for expanded agricultural land are tempting, and China has plans for hundreds more dams, there are now many well-documented environmental impacts that need to be considered at the earliest stages in any dam planning process. These can no longer be swept under the carpet or dismissed. They are real, they are inevitable, and they need to be weighed as costs, just like the price of concrete and steel.

DAM REMOVAL

There is no question that dams provide important benefits and also no argument that they come with some significant and long-term effects. Dams globally are aging, and there are now about five thousand large dams around the world that are over seventy years old. As a result, some of the important benefits they initially provided have been reduced or eliminated altogether. One of the major factors has been sediment accumulation. Many reservoirs have trapped so much sediment that their potential for flood control and their capacity to store water are only a fraction of their original capacity (figure 17.4). Recreational uses have also been nearly eliminated in many places as sediment and vegetation have replaced water. Compounding these losses, sediment is heavier and places a greater load on a dam than water and, combined with newer data on the intensity of seismic shaking in earthquake-prone areas like California, China, Turkey, and India, the safety of many dams is now a growing concern.

For decades, dam removal has been an attractive approach for dealing with obsolete, unsafe, or unwanted dams. Many older dams have completely outlived their original purpose and no longer provide any benefits. The thousands of small dams that were originally built to power mills that began the industrial revolution in the eastern United States come to mind. The mills are long gone, but the dams are still there and in many cases have become useless liabilities.

While taking down a dam is not without its concerns, challenges, and costs, there have now been hundreds of small dams that have been removed in the United States. Removal of a dam generates all of the benefits that any free-flowing river provides naturally. Fish like salmon, steelhead, and shad can return to their original spawning grounds and expand their populations. Sediment, whether sand, silt, or clay, can move downstream to ultimately nourish floodplains, deltas, and

FIGURE 17.7. The Elwha River flowing through the site of former Glines Canyon Dam, Olympic Peninsula, Washington (Photo: Jeff Duda, U.S. Geological Survey)

beaches. Wildlife can return to formerly flooded areas, and people can enjoy the benefits of free-flowing streams, whether fishing, canoeing, or rafting. Between 1990 and 2015 an estimated 900 dams were removed in the United States, and about 50 to 60 additional dams are now being removed each year. Although many of the early dams removed were small, this has changed in recent years as the safety, migratory fish, and sediment issues have become more apparent and experience has been gained regarding the removal processes, environmental impact assessment, and the required government agency and stakeholder involvement. While dam removal is not appropriate for most of the nation's or the world's very large dams, there are many good candidates being considered, important lessons that have been learned, and some encouraging success stories.

After more than a decade of study and planning, the largest dam removal project in U.S. history was completed in 2014 on the Elwha River on Washington's Olympic Peninsula. This $350 million project, driven largely by restoration of river habitat and salmon migration, removed the 108-foot-high Elwha Dam and the 201-foot-high Glines Canyon Dam (figure 17.7). Dam removal allowed salmon and steelhead to travel upriver past the dam sites, an event that had not occurred since the first dam was built in 1913, when salmon runs of 400,000 fish annually had taken place. After spawning, the salmon die and their carcasses decompose, releasing nutrients to nourish downstream ecosystems. Since the dam removal, reservoir beds that looked like moonscapes have quickly returned to rich vibrant habitats.

Another recent and significant dam removal was that of the San Clemente Dam on Central California's Carmel River. This 106-foot-high structure, originally built for water supply for the semiarid Monterey Peninsula, had been almost completely filled with sediment and had been deemed unsafe in 1992 (see figure 17.4). This $83 million, two-year project was completed in 2015 and now allows sand to be transported downstream to the shoreline to nourish beaches, and steelhead to migrate upstream to spawn.

Several additional dams in California are next on the list for removal and awaiting funding. The Matilija Dam on the Ventura River blocks steelhead spawning grounds and is now approximately 90 percent full of sediment, including 3 million cubic yards of sand that could nourish the eroding shoreline at the river mouth. Fifty miles south, the privately owned Rindge Dam, on Malibu Creek, was built in 1924 and is now virtually filled with sediment, making it both unsafe and obsolete. Proposals have been considered for years for removing the dam and allowing the impounded sand to be transported downstream to the shoreline where beaches could again be nourished.

Many more projects are being discussed and have been proposed. In each case there are complex issues to resolve, including costs, environmental approvals, positive and negative biological impacts, and the best approach for dam removal and getting impounded sediment downstream to the shoreline. Each completed project provides additional experience and perspective and should, over time, make future projects easier, more efficient, and more effective.

## COASTAL ARMORING AND SAND SUPPLY

Coastal bluffs or cliffs that consist of sand-sized material and that are retreating at significant rates provide some sand to the adjacent beaches. The amount of sand contributed from any particular section of coastal bluff can be determined or calculated if you know (1) the alongshore length and height of the bluff; (2) the percentage of beach-compatible sand contained in the bluff materials; and (3) the average annual or long-term bluff retreat rate. Depending on other sources of beach sand within the region or littoral cell, the volume contributed by bluff retreat may be

FIGURE 17.8. Seawalls or revetments have now been built along 33 percent of the shoreline of Southern California, eliminating the addition of sand to beaches from bluff erosion. (Photo Kenneth and Gabrielle Adelman © 2002, California Coastal Records Project, www.Californiacoastline.org)

very important. Where there are very large volumes of sand delivered by other sources, such as rivers or streams, the bluff input may be relatively minor. For most of California's littoral cells, cliff or bluff erosion is only a modest contributor to the beach sand budgets, averaging about 10 percent but as high as 20 or 30 percent in some areas. The importance of bluff erosion as a source of beach sand is generally higher in Southern California because the rivers and streams are smaller and often are intermittent in their flow due to low annual rainfall, and many of them have multiple dams that trap sand before it reaches the shoreline.

Nonetheless, where bluff erosion is a significant sand contributor to the fronting beach and those bluffs are proposed for armoring, then this sand supply will be cut off or reduced. Thirty-three percent of the entire Southern California coastline has now been armored with seawalls or riprap revetments (figure 17.8), which has eliminated about 35,000 cubic yards of sand from the beaches each year. It is relatively straightforward to determine the amount of sand that will no longer be contributed to the local beaches due to armoring and whether or not this will have a significant regional impact. In California, if the potential loss is significant, a sand mitigation fee becomes part of the seawall permit. If used to replace lost sand, this is one approach for helping to make up the loss.

BEACH SAND MINING: A GLOBAL ISSUE

We live in a sand world, but very few of us realize this. Our lives are totally surrounded by and dependent on this simple material. Sand is everywhere. It's used in plastic, metal alloys, paints, cosmetics, toothpaste, detergents, paper, glass, and computer chips, and it's a key ingredient in concrete, which is the world's most widely used construction material. Streets and highways, bridges and airports, sidewalks and skyscrapers, all contain large amounts of sand. In fact, after water, sand is the most consumed resource on the planet.

The concrete in a typical home requires about 200 tons of sand; a large building like a hospital, perhaps 2,000 tons and a mile of highway, 30,000 tons. A large nuclear power plant with its massive cooling towers can require up to 12 million tons of sand. And we continue to build a lot of each of these. The combined use of all sand and gravel around the world (construction, land reclamation and shoreline developments, and industry products) totals approximately 44 billion tons annually. To get a clearer idea of how much concrete the world pours each year in new construction, if we were to instead put it all into a wall extending completely around the Earth at the equator, it would be 88 feet wide and 88 feet high.

Few people would probably view sand as a limited resource or consider that there would be any concerns associated with mining, quarrying, or extracting it. But there are problems, global problems, and they are getting worse, simply because of the seemingly insatiable appetite of modern civilization for sand; and good sand is becoming harder to come by. For decades, sand quarries and riverbeds were the sources for most of our construction sand, but many of the easy to access deposits have now been used up, and there are now environmental concerns with dredging up riparian corridors.

Our demand for sand has also increased astronomically for two reasons. In addition to all of the products listed above, the world of electronics has grown quickly, whether computers, cell phones, or any of our other communication and computing tools. These devices all require silicon dioxide. And our source for silicon dioxide is quartz, typically the most common mineral in sand.

The second reason is the massive amount of construction that is accompanying population growth and global development. Dubai is a good example of this trend. It had the fastest growing economy in the world from 1975 through 2008, expanding by a factor of 11. The population

FIGURE 17.9. Crescent Island at Palm Jumeirah in Dubai, which has been artificially created by dredging and depositing offshore sand. (Photo: by Citizen59 licensed under CC BY-SA 2.0 via Flickr)

nearly doubled in the nine years between 2007 and 2016, growing from 1.8 million to 3.5 million people. This extremely rapid growth, particularly evident in its high-rise skyline, has literally risen from the desert. But this development has required hundreds of millions of tons of sand in the process, either for use in concrete for roads and buildings (30 percent of Dubai's office space is now vacant, however) or for creating the widely publicized artificial islands (the World and the Palms; figure 17.9) that were created offshore for proposed new development. The Palm Jumeirah Island alone required 425 million tons of sand. But Dubai's usable sand for concrete has run out. In order to build the new World Archipelago and the new Burj Khalifa tower, just over half a mile high and the world's tallest building, they had to import sand from Australia, thousands of miles away. In 2012 neighboring Qatar, was reportedly the world's largest importer of sand for construction, some $6.5 billion worth. This may all seem surprising for countries like Qatar and the United Arab Emirates that are nearly covered with desert sand. Windblown sand, however, is typically much finer grained and very rounded and does not bind well in making concrete, so it has not typically been used in construction.

There is a reason we use so much concrete: it's a great building material. The production of cement, however, which is a key ingredient of concrete, has a very large global carbon footprint because we use so much of it, and the amount continues to grow. Making cement produces massive quantities of carbon dioxide. About 5 percent of all the anthropogenic global $CO_2$ emissions come from producing cement. China alone accounts for about 60 percent.

FIGURE 17.10. This sand mining operation along the shoreline of southern Monterey Bay is the only site in the United States where sand is being mined directly from the beach. The removal of very large volumes of sand for decades has led to significant shoreline erosion problems. (Photo: Kenneth and Gabrielle Adelman © 2005, California Coastal Records Project, www.Californiacoastline.org)

Another key component of concrete is sand, lots of sand. A typical concrete mixture consists of 1 part cement, 2 parts sand, and 3 parts aggregate or gravel. And today much of that sand is coming from beaches or the shallow seafloor just offshore. That sand is used not only for concrete but also for construction fill, creating new land (or islands in Dubai's case), and nourishment of other beaches. No matter where it takes place, the direct removal of beach sand is usually destructive (figure 17.10). In places like South Asia and Africa, removal of beach sand is typically carried out with no management or controls, and more often than not illegally. Beach or nearshore sand mining is now a widespread global phenomenon and has been documented on five continents and in seventy-three countries, including Morocco, South Africa, India, Cambodia, Indonesia, and Malaysia and several Caribbean islands. Why beach sand? Because it is generally very easy to obtain, and in many places, ownership is not that clear. It is the Tragedy of the Commons all over again, this time along the shoreline.

Singapore has been an amazing economic success story, but being a small city-state, its growth was historically constrained by its land area.

Not to be limited by this restriction, Singapore decided to just expand its edges. Over the past fifty years it has increased its land area by 22 percent, in large part by importing enormous volumes of sand from the beaches of Indonesia, Malaysia, and Cambodia (see figure 1.5). Ninety-five percent of Singapore's sand has been imported since 1995, some 260 million tons, making it the largest global importer of sand, but not without effects at the dredging sites. The sand mining was reported to be responsible for the disappearance of twenty-four Indonesian sand islands. As demand has increased in Singapore to further expand build-able area, the price of sand has escalated. Between 1995 and 2001 sand was being imported at about $3 per ton, but between 2003 and 2005 the price escalated to $175 per ton. As other governments banned the extraction of their beach sand, the mining and importing continued but often illegally. It is also now widely reported that because of the value of this increasingly scarce but critical construction resource, sand mafias have emerged in India, Southeast Asia, and North Africa, which now control the trade in sand. It's a difficult commodity to track.

This theft of beach sand, in addition to complete removal of sand islands, is also a direct cause of erosion along many shorelines. In Morocco, which is an extreme example, beaches have been completely removed (figure 17.11a, b). Ironically, here and elsewhere, the beaches are being eliminated and the sand is going into the concrete used to build the high-rise hotels or apartments to accommodate the people who come to enjoy them.

Wide beaches provide the best natural protection from the storm surges associated with tropical cyclones and hurricanes, from high tides combined with storm waves, and even from tsunamis. Beach sand mining takes away that protection and exposes people and their homes, infrastructure, and other development to the forces of storms and other coastal hazards. At a time when sea level is rising globally at an increasing rate, coasts everywhere need all the protection they can get, and removing or narrowing the beach only increases the risks and exposure (figure 17.12).

## THE FUTURE OF BEACHES AND BEACH SAND: WHERE DO WE GO FROM HERE?

Sand has become a nonrenewable resource, as the beaches in many areas of the world can attest. We are using it up, in most cases locking it into concrete, faster than natural processes of erosion and transport

FIGURE 17.11a, b. Illegal beach sand mining in South Tangier, Morocco. (Photo: © SAF—Coastalcare; www.coastalcare.org)

FIGURE 17.12. This beach in Malibu, California, has narrowed and nearly disappeared as a result of a combination of upcoast sand reduction and placement of protective structures. (Photo: D. Shrestha Ross © 2008)

can replace it. Sustainability of our sand supplies is a critical objective to be achieved if we are to maintain the world's beaches and the recreation and tourism industries they support, as well as the protection they provide. Three common approaches for other limited resources can help accomplish this goal: reuse, reduce, and recycle. One important approach is to optimize the use of existing buildings instead of simply demolishing them to make way for newer structures. Too often we tear down older buildings without considering the potential for new ways of using them instead of utilizing more natural resources, like sand, for the needed concrete. The same principle applies to highways and roads. In some cases, concrete or asphalt can be broken up, recycled, and used as aggregate in new highways or infrastructure. Switzerland, for example, has been subsidizing use of aggregate that incorporates recycled concrete for years. There are also the millions of tons of sand and gravel that have been trapped at the bottom of reservoirs behind the world's dams that could be utilized. Removal of these sediments is not simple, has costs, but would produce multiple benefits. There are also large deposits of sand and gravel deposited on the broad floodplains of many large rivers, which are well away from the active channels and riparian corridors, that could be utilized.

Some substitutes for sand in concrete have been developed, and more research needs to be conducted into other aggregate alternatives. Use of quarry dust, incinerator ash, and even desert sand, if mixed with other materials, have all proven feasible. Although there are alternative building materials for houses, when it comes to large buildings, freeways, roads, and parking garages and many other larger structures, concrete is the logical and favored material, and the construction industry is geared toward building with concrete. This presents some major challenges to engineers, material scientists, architects, and builders.

Sand is the natural resource used in larger volume than any other raw material on Earth except water, and it is now being utilized at a greater rate than it is being renewed. Rapid economic growth, particularly in Asia, is increasing the demand for sand for construction, and the negative environmental impacts have become obvious in many coastal areas. In contrast to many other environmental issues, there is a much larger gap between the magnitude of the problem and the public perception of it, such that action to date has been minimal. Some governments have taken a strong stance and outlawed the mining of beach sand, and these policies need to be exported to other nations facing similar losses. Illegal sand mining can be halted, but it takes political will and responsible politicians who are not connected to the sand mining industry and who understand the multiple impacts of the loss of beaches.

# Index